U.S.NRC

United States Nuclear Regulatory Commission

Protecting People and the Environment

NUREG-1482, Rev. 2

Guidelines for Inservice Testing at Nuclear Power Plants

Inservice Testing of Pumps and Valves and Inservice Examination and Testing of Dynamic Restraints (Snubbers) at Nuclear Power Plants

Final Report

Office of Nuclear Reactor Regulation

AVAILABILITY OF REFERENCE MATERIALS
IN NRC PUBLICATIONS

United States Nuclear Regulatory Commission

Protecting People and the Environment

Guidelines for Inservice Testing at Nuclear Power Plants

Inservice Testing of Pumps and Valves and Inservice Examination and Testing of Dynamic Restraints (Snubbers) at Nuclear Power Plants

Final Report

Manuscript Completed: September 2013
Date Published: October 2013

Prepared by:
Gurjendra S.Bedi

Office of Nuclear Reactor Regulation

ABSTRACT

The staff of the U.S. Nuclear Regulatory Commission (NRC) is issuing Revision 2 to the NUREG-1482, "Guidelines for Inservice Testing at Nuclear Power Plants," to assist the nuclear industry in establishing a basic understanding of the regulatory basis for pump and valve inservice testing (IST) programs and dynamic restraint (snubbers) examination and testing programs. This NUREG also provides information regarding the NRC's involvement in the development of the American Society of Mechanical Engineers (ASME) *Code for Operation and Maintenance of Nuclear Power Plants* (OM Code). In this NUREG, the staff discusses OM Code inquiries, the inservice examination and testing of snubbers, pump and valve IST, the use of ASME code cases, conditions on the use of the OM Code, guidance for OM Code noncompliance, requests for alternatives to the OM Code at operating commercial nuclear power plants, and the development of IST programs for new reactors*.

This NUREG report replaces Revision 0 and Revision 1 to NUREG-1482 and is applicable, unless stated otherwise, to editions and addenda (up to the 2004 Edition including 2005 and 2006 addendas) to the ASME OM Code, that are incorporated by reference in Title 10 of the *Code of Federal Regulations* (10 CFR) 50.55a(b). In addition, the staff discusses other IST program topics such as the NRC process for the review of the OM Code, conditions on the use of the OM Code, interpretations of the OM Code, and development of IST programs for new reactors. In this NUREG, the staff provides guidance included in Revision 1 to NUREG-1482 that has been updated to reflect IST lessons learned and operating experience since the NUREG was previously issued. A new Appendix A contains guidance related to inservice examination and testing of snubbers.

PAPERWORK REDUCTION ACT STATEMENT

PUBLIC PROTECTION NOTIFICATION

Note: *New Reactor is defined as a nuclear plant that was issued (or will be issued) its construction permit, or combined license for construction and operation, by the applicable regulatory authority on or following January 1, 2000

CONTENTS

Figures

Tables

EXECUTIVE SUMMARY

The information in NUREG-1482, Revision 0, issued April 1995, and Revision 1, issued January 2005, described IST programs in the past. Revision 2 to NUREG-1482 replaces Revision 0 and Revision 1 to NUREG-1482 and is applicable, unless stated otherwise, to editions and addenda (up to the 2004 Edition including 2005 and 2006 addendas) to the ASME OM Code, that are incorporated by reference in Title 10 of the *Code of Federal Regulations* (10 CFR) 50.55a(b) (Federal Register, Vol. 76, No. 119, page 36232-36279, dated June 21, 2011).

This NUREG provides guidance for the inservice testing of pumps and valves, and inservice testing of dynamic restraints (snubbers) at nuclear power plants based on lessons learned since the issuance of Revision 0 and Revision 1 to NUREG-1482. This NUREG contains guidance provided in Revision 1 to NUREG-1482 for pumps and valves that has been updated for the development of IST programs at nuclear power plants. A new Appendix A have been added which contains only guidance related to inservice examination and testing of snubbers,

The guidelines and recommendations provided in this NUREG and its Appendix A do not supersede the regulatory requirements specified in 10 CFR 50.55a, "Codes and standards." Further, this NUREG and its Appendix A do not authorize the use of alternatives to, or grant relief from, the ASME Code requirements for inservice testing of pumps and valves, or inservice examination and testing of dynamic restraints (snubbers), incorporated by reference in 10 CFR 50.55a.

ABBREVIATIONS

ADAMS	Agencywide Documents Access and Management System
ADS	automatic depressurization system
AFW	auxiliary feedwater
ALARA	as low as reasonably achievable
ANSI	American National Standards Institute
AOV	air-operated valve
ASME	American Society of Mechanical Engineers
BEP	best efficiency point
BWR	boiling-water reactor
BWST	borated water storage tank
CFR	*Code of Federal Regulations*
COL	combined operating license
CPT	comprehensive pump test
CRD	control rod drive
CVCS	chemical and volume control system
DBD	design-basis document
ECCS	emergency core cooling system
FR	Federal Register
FSAR	final safety analysis report
GDC	general design criterion
GE	General Electric Company
GL	generic letter
GPM	gallons per minute
GSI	generic safety issue
HCU	hydraulic control unit
HOV	hydraulic-operated valve
HPCI	high-pressure coolant injection
Hz	hertz
IEEE	Institute of Electrical and Electronics Engineers
IM	inspection manual
IN	information notice
IP	inspection procedure
ISI	inservice inspection
IST	inservice testing
ITAAC	inspections, tests, analyses, and acceptance criteria
JOG	Joint Owners Group

LCO	limiting condition for operation
LOCA	loss-of-coolant accident
LWR	light-water reactor
MOV	motor-operated valve
MSIV	main steam isolation valve
MSSV	main steam safety valve
NEI	Nuclear Energy Institute (formerly NUMARC)
NIC	Nuclear Industry Check Valve Group
NOED	Notice of Enforcement Discretion
NRC	U.S. Nuclear Regulatory Commission
NRR	Office of Nuclear Reactor Regulation (NRC)
NUMARC	Nuclear Management and Resources Council (now NEI)
OM Code	ASME *Code for Operations and Maintenance of Nuclear Power Plants*
PASS	post-accident sampling system
PdM	predictive maintenance
PIV	pressure isolation valve
PORV	power-operated relief valve
POV	power-operated valve
PRA	probabilistic risk assessment
PTC	Performance Test Code
Psid	pounds per square inch differential
PST	pre-service testing
PWR	pressurized-water reactor
P&ID	piping and instrument diagram
RCIC	reactor core isolation cooling
RCPB	reactor coolant pressure boundary
RCS	reactor coolant system
RG	regulatory guide
RHR	residual heat removal
RIS	regulatory issue summary
RTNSS	regulatory treatment of nonsafety systems
RWST	refueling water storage tank
RWT	refueling water tank
SAR	safety analysis report
SBLC	standby liquid control
SGCV	Code Committee Sub-Group on Check Valves (ASME)
SE	safety evaluation
SI	safety injection
SOV	solenoid-operated valve
SR	surveillance requirement
SRP	Standard Review Plan
SSC	system, structure, and/or component

S/RV	safety/relief valve
STS	Standard Technical Specifications
TRM	technical requirement manual
TS	technical specification(s)
UFSAR	updated final safety analysis report
WGCV	Working Group on Check Valves (ASME)

PREFACE

NUREG publications consist of reports or brochures on regulatory decisions, results of research, results of incident investigations, and other technical and administrative information. Some of the information herein is similar in appearance to U.S. Nuclear Regulatory Commission (NRC) staff positions given in a regulatory guide.

1. INTRODUCTION.

1.1   Regulatory Basis

Title 10, Section 50.55a, "Codes and standards," of the *Code of Federal Regulations* (10 CFR 50.55a) defines the requirements for applying industry codes and standards to boiling-or pressurized-water-cooled nuclear power facilities. Each of these facilities is subject to the conditions in paragraphs (a), (f), and (g) of 10 CFR 50.55a, as they relate to inservice inspection (ISI) and inservice testing (IST). By rulemaking effective September 8, 1992 (*see* 57 FR 34666; August 6, 1992), the U.S. Nuclear Regulatory Commission (NRC) established paragraph (f) of 10 CFR 50.55a to separate the IST requirements from the ISI requirements in paragraph (g).

The regulations at 10 CFR 50.55a, "Codes and standards," define the requirements for applying industry codes and standards to boiling- or pressurized-water-cooled nuclear power facilities. The National Technology Transfer and Advancement Act of 1995 (P.L. 104-113) requires that if agencies establish technical standards, they must use technical standards that voluntary consensus standards bodies develop or adopt unless the use of such standards is inconsistent with applicable law or is otherwise impractical. P.L. 104-113 requires Federal agencies to use industry consensus standards to the extent practical; however, it does not require Federal agencies to endorse a standard in its entirety. The law does not prohibit an agency from generally adopting a voluntary consensus standard while taking exception to specific portions of the standard if those provisions are deemed to be "inconsistent with applicable law or otherwise impractical." Furthermore, taking specific exceptions furthers the congressional intent of Federal reliance on voluntary consensus standards because it allows the adoption of substantial portions of consensus standards without the need to reject the standards in their entirety because of limited provisions that are not acceptable to the agency.

The American Society of Mechanical Engineers (ASME) *Code for Operation and Maintenance of Nuclear Power Plants* (OM Code) is a national, voluntary consensus standard. The NRC approves or mandates the use of editions and addenda to the codes in 10 CFR 50.55a through the rulemaking process of "incorporation by reference." Once the ASME Code Edition or Addenda is incorporated by reference into the NRC's regulations, each provision of the codes that 10 CFR 50.55a incorporates by reference and mandates constitutes a legally binding NRC requirement imposed by rule.

As of July 21, 2011, the NRC regulations in 10 CFR 50.55a(b)(3) incorporate by reference the 1995 Edition through the 2004 Edition with 2005 and 2006 addenda of the ASME OM Code promulgated by the ASME, in which Subsection ISTA provides general IST requirements and Subsections ISTB, ISTC, and ISTD provide the IST requirements for pumps, valves, and dynamic restraints, respectively. Based on those requirements, each of the NRC's nuclear power plant licensees must establish IST programs, specify the components included in the program as well as the test methods and frequencies for those components, and implement the program in accordance with the OM Code.

Where a test requirement of the OM Code is determined to be impractical for a facility, the NRC's regulations in 10 CFR 50.55a(f)(6)(i) allow the licensee to submit a request for relief from the given requirement, along with information to support the determination. Relief requests generally detail the reasons for deviating from the Code requirements and propose alternative testing methods or frequencies. The Commission is authorized to evaluate licensees' relief requests, and may grant the requested relief or impose alternative requirements, considering the burden that the licensee might incur if the Code requirements were enforced for the given facility. Pursuant to 10 CFR 50.55a(a)(3)(i) and (ii), the Commission may also authorize the licensee to implement an alternative to the Code requirements, provided that the alternative ensures an acceptable level of quality and safety or the Code requirement presents a hardship without a compensating increase in the level of quality and safety.

The regulations in 10 CFR 50.55a(f)(4)(iv) specify that inservice testing of pumps and valves may meet the requirements in editions and addenda of the OM Code that were published more recently than those that are incorporated by reference in 10 CFR 50.55a(b), subject to Commission approval and the limitations and modifications listed in 10 CFR 50.55a(b). Requests for approval to use later editions and addenda previously incorporated by reference in 10 CFR 50.55a may be made via letter to the NRC. See NRC Regulatory Issue Summary (RIS) 2004-12, "Clarification on Use of Later Editions and Addenda to the ASME OM Code and Section XI," for clarification.

The 10 CFR 50.55a regulations are applicable to pump and valve IST programs at operating reactors and are discussed throughout this NUREG. As a result of the unique wording in various paragraphs, note that the NRC **authorizes** licensee-proposed alternatives in accordance with 10 CFR 50.55a(a)(3), **grants** relief and **imposes** alternative requirements in accordance with 10 CFR 50.55a(f)(6)(i) and 10 CFR 50.55a(g)(6)(i), or **approves** the use of later code editions and addenda in accordance with 10 CFR 50.55a(f)(4)(iv) and 10 CFR 50.55a(g)(4)(iv).

1.2   Regulatory History of Staff Guidance on IST

The NRC previously issued guidance for implementing IST requirements. After publishing the rule that established the IST requirements in § 50.55a (see 41 FR 6256; February 12, 1976), the NRC sent letters to notify operating licensees of the new rule. In November 1976, after receiving inquires from the licensees regarding acceptable methods for complying with the regulation, the NRC issued letters to licensees to transmit Staff Guidance for Complying with 10 CFR 50.55a(g), "Inservice Inspection Requirements."

To eliminate the backlog of IST program reviews for operating nuclear power plants, the NRC issued Generic Letter (GL) 89-04, "Guidance on Developing Acceptable Inservice Testing Programs," dated April 3, 1989 (Agencywide Documents Access and Management System (ADAMS) Accession No. ML031150259). That generic letter included 11 technical positions that the staff uses in reviewing licensees' IST program relief requests and described alternatives to the Code requirements that the staff considered acceptable. As a unique resolution of the backlogged IST program reviews, in GL 89-04, the staff also approved six of the 11 technical positions (1, 2, 6, 7, 9, and 10), pursuant to 10 CFR 50.55a(g)(6)(i), with the provision that the licensee must perform the alternative testing delineated in the applicable position. The staff approved these alternatives upon recognizing that it might be impractical to perform the required testing, and enforcing the requirements might pose an unnecessary burden on licensees.

GL 89-04 stated that licensees must document their use of Positions 1, 2, 6, 7, 9, and 10 in the IST program, but did not require the documentation to take the form of a relief request. The generic letter granted approval to follow the alternative testing delineated in Positions 1, 2, 6, 7, 9, and 10, pursuant to 10 CFR 50.55a(g) [now (f)] provided that the licensee followed the provisions of GL 89-04. For convenience, most licensees documented their use of these positions in relief requests; however, the NRC found other forms of program documentation were acceptable as long the provisions of the referenced positions were clearly documented and were discussed in adequate detail to ensure that they conformed with those provisions. Certain licensees may have submitted relief requests to ensure they adequately documented their conformance in their IST program, even though documentation in the program also would have been acceptable, as stated in GL 89-04.

The staff held four public meetings to discuss GL 89-04 and stated that the generic letter was a first step toward resolving various problems associated with developing and implementing IST programs at nuclear power plants. The staff had previously identified these problems through its reviews of licensees' IST programs, and by inspecting and auditing IST programs at plant sites, participating on the ASME Code committees, and meeting with licensees and industry groups. The staff then summarized the questions and answers from the four public meetings in a letter entitled "Minutes of the Public Meetings on Generic Letter 89-04," dated October 25, 1989, (ADAMS Legacy Accession No. 9001040128). That letter contained useful information about how to apply the guidance in GL 89-04 and discussed issues of interest to licensees who attended the public meetings. In a subsequent letter, dated September 16, 1991, the staff issued "Supplement to Minutes of the Public Meetings on Generic Letter 89-04" to address a question on stop-check valve testing.

Since the NRC issued GL 89-04, the staff has improved its guidance regarding IST by revising 10 CFR 50.55a and separating the IST and ISI programs in paragraphs (f) and (g), respectively, issuing additional guidance, and coordinating with ASME for periodic symposia on testing pumps and valves. The NRC intends to continue to improve its IST-related guidance through continued participation in Code and technical organizations, as well as regular updates of the agency's published guidance as future needs arise.

1.3   NRC Recommendations and Guidance

The NRC staff guidance and recommendations in NUREG-1482, Revision 2, "Guidance for Inservice Testing at Nuclear Power Plants," are based on ASME OM Code, 1998 Edition through the 2004 Edition with 2005 and 2006 addenda.

The recommendations herein replace the guidance and technical positions in GL 89-04. Note that specific relief is required to implement the guidance derived from GL 89-04. However, relief justification may refer to the positions in the GL with clarifying information to clearly show how it would apply to a licensee's situation. To the extent practical, this document reflects the applicable section, subsection, or paragraph of the appropriate documents (subsections of 10 CFR Part 50, "Domestic Licensing of Production and Utilization Facilities"; OM Code; regulatory guides; etc.).

The guidance presented herein is voluntary and may be used for requesting relief under 10 CFR 50.55a(f), or for approval of an alternative under 10 CFR 50.55a(a)(3). Licensees may also request relief or the use of an alternative which is not in conformance with this guidance. The NRC may grant relief or authorize the alternative if the licensee has addressed all of the aspects of the relief or alternative is an acceptable manner.

1.3.1 NRC Review of the ASME OM Code

The first edition and addendum to the OM Code that 10 CFR 50.55a incorporated by reference were the 1995 Edition and the 1996 Addendum. The NRC determines acceptability of new provisions in new editions and addenda to the OM Code and the need for conditions on the use of the OM Code. Generally, the NRC staff participates with other ASME committee members in discussions and technical debates in the development of OM Code revisions. NRC committee representatives discuss the codes and technical justifications with other cognizant NRC staff to ensure an adequate technical review. Finally, NRC management reviews and approves the proposed agency position on the OM Code as part of the rulemaking to amend 10 CFR 50.55a to incorporate by reference new editions and addenda to the OM Code and conditions on its use. This process, when considered along with ASME's own process for developing and approving the OM Code, provides reasonable assurance that the NRC approves for use only those new and revised OM Code editions and addenda (with conditions as necessary), that provide reasonable assurance of adequate protection to public health and safety and that do not have significant adverse impacts on the environment.

1.3.2 Exemptions

Under 10 CFR 50.12(a) and 10 CFR 52.7, the NRC may, either on its own initiative or upon application by any licensee, grant an exemption from the requirements of 10 CFR Part 50 that is authorized by law, does not present an undue risk to the public health and safety, is consistent with the common defense and security, and is appropriate because of special circumstances. If the NRC approves the application, the exemption relieves the licensee from compliance with the regulation(s) involved. Exemptions are normally not used for 10 CFR 50.55a(a)(3) approval of alternatives, or § 50.55a(f) reliefs.

1.4 Synopsis of Report

This appendix follows the format of a typical IST program plan, including Development and Implementation, General Guidance, Valves, Pumps, Technical Specifications, Code Noncompliance, and Risk-Informed IST.

Section 2, "Developing and Implementing an IST Program," describes existing IST requirements, discusses the scope of the IST program, and describes guidance for presenting information in IST programs, including cold shutdown justifications, refueling outage justifications, and relief requests. Section 2 also includes a sample list of plant systems for boiling-water reactors (BWRs) and pressurized-water reactors (PWRs) that typically (but not necessarily) contain pumps or valves that perform a safety function and are subject to requirements of the OM Code.

Section 3, "General Guidance on Inservice Testing," describes the NRC's recommendations and its bases for several general aspects of IST. Sections 4 and 5 then become more specific, describing recommendations on valve-related and pump-related issues, respectively. Throughout Sections 3 through 5, this document discusses the IST requirements for which licensees have requested relief or proposed alternatives. This document also provides guidance concerning the types of information that licensees typically should (or in some cases must) include in their relief requests. Sections 3 through 5 also discuss related Code and regulatory issues and provide recommendations and guidance as needed. These discussions do not impose additional requirements beyond those imposed by the Code or the regulations and, as such, do not represent backfits.

These discussions are intended to clarify the existing requirements of the Code or the regulations and, as such, they may provide recommendations to ensure that licensees continue to meet the Code and other regulatory requirements.

Sections 6, 7, and 8 discuss the standard technical specifications, the process licensees should follow when they identify a Code nonconformance, and the development of a risk-informed IST program. Section 9 presents a list of related references.

This guidance is not equivalent to staff positions in a regulatory guide, because licensees must request approval through the relief or alternative process described in 10 CFR 50.55a where the ASME OM Code will not be met.

2. DEVELOPING AND IMPLEMENTING AN INSERVICE TESTING PROGRAM

Licensees may use the following guidance for developing and implementing inservice testing (IST) programs. This guidance supplements existing requirements and previously approved guidance on IST.

2.1   Compliance Considerations

The NRC regulations in 10 CFR 50.55a specify requirements for IST of certain safety-related pumps and valves that must be tested according to the requirements of the ASME OM Code. This testing is intended to assess the operational readiness of the stated components. Specifically, the regulations state that the tests conducted during the initial and successive 120-month intervals must be based on the requirements in the applicable edition and addenda of the ASME OM Code, to the extent practical, within the limitations of design, geometry, and materials of construction, as described in 10 CFR 50.55a(f)(4).

In addition, Paragraph 50.55a(f)(4)(ii) requires that IST conducted during each 120-month interval following the initial interval must be conducted in compliance with the requirements of the latest edition and addenda of the Code incorporated by reference in the version of 10 CFR 50.55a(b) that is in effect 12 months before the start of the interval. As of July 21, 2011, the NRC regulations in 10 CFR 50.55a(b) incorporate by reference the 1995 Edition through the 2004 Edition including 2005 and 2006 addenda of the ASME OM Code subject to conditions.

The regulations specify the requirements for IST, and the requirements of the OM Code, as incorporated by reference into the regulations, and therefore are legally-binding requirements. A plant's technical specification (TS) which include general and specific requirements for IST and other surveillance testing of pumps and valves, are part of the license and therefore are legally-binding requirements. The plant's safety analysis includes information concerning the design limitations and functional requirements for the performance of pumps and valves for the given facility. The plant's IST program, including any relief requests and data analysis methods, describes the licensee's means for implementing the various requirements for the specific plant.

The implementing procedures include the lowest tier of IST elements. In addition, IST engineers often use other information (such as bases documents, vendor manuals, trend data, and graphs) in developing, maintaining, and implementing the plant's IST program.

Licensees must meet the regulations and TS if they identify a conflict between the regulations and the licensees' program or procedures. The staff gives guidance on cases where a licensee modifies its plant in a way that affects the basis for relief that the NRC has previously granted. Similarly, if a licensee has obtained the NRC's approval of an alternative pursuant to 10 CFR 50.55a(a)(3)(i) or (ii), the licensee need not use that alternative if it subsequently determines that continued compliance with the Code requirements is warranted or necessary for particular circumstances that may preclude implementation of the approved alternative. When a licensee revises an implementing procedure, the licensee typically ensures that the IST program continues to reflect the required testing. Similarly, when a system, subsystem, or component is modified, or an operating or test procedure or valve alignment is changed in accordance with

10 CFR 50.59, "Changes, Tests, and Experiment," the licensee typically reviews the IST requirements to determine whether it must change the program for the affected components.

The NRC may authorize alternatives to Code testing requirements submitted as relief requests or in a similar format that include a discussion of the requirements, a description of the proposed alternative, and the justification for approval of the alternative. See 10 CFR 50.55a the following provisions for accepting alternatives or granting relief:

- Regulations in 10 CFR 50.55a(a)(3)(i) allow the NRC to authorize alternatives if the proposed alternatives would provide an acceptable level of quality and safety. The NRC will normally authorize an alternative pursuant to this provision only if the licensee proposes a method of testing that is equivalent to, or an improvement of, the method specified by the Code, or if the testing will comply or is consistent with later Code editions approved by the NRC in 10 CFR 50.55a(b).

- Regulations in 10 CFR 50.55a(a)(3)(ii) allow the NRC to authorize an alternative if compliance [with the Code requirement] would result in hardship or unusual difficulty without a compensating increase in the level of quality and safety. The NRC may authorize an alternative pursuant to this provision if, although the proposed alternative testing does not comply with the Code, the increase in overall plant safety and quality attained by complying with the Code requirement is not justified in light of the difficulty associated with compliance.

- Regulations in 10 CFR 50.55a(f)(6)(i) include the following provision:

  The Commission will evaluate determinations that Code requirements are impractical. The Commission may grant relief and may impose such alternative requirements as it determines is authorized by law, giving due consideration to the burden upon the licensee that could result if the requirements were imposed on the facility.

The NRC may grant relief pursuant to this provision or may authorize alternatives if the licensee demonstrates that the design or access limitations make the Code requirement impractical. Thus, the staff's evaluation considers the burden created by imposing the Code requirements on the licensee.

2.1.1 ASME Code Case Applicability

A Code Case is the official method of the ASME for handling a reply to an inquiry when study indicates that the Code wording needs clarification, or when the reply modifies the existing requirements of the Code, or grants permission to use alternative methods. ASME develops Code Cases through a consensus process to clarify the intent of existing Code requirements or to provide an alternative to a specific Code requirement. A Code Case may be issued for the purpose of providing alternative rules when justified, to permit early implementation of an approved revision when the need is urgent, or to provide rules not covered by existing provisions of the ASME OM Code.

The NRC reviews new or revised Code Cases to determine their acceptability for incorporation by reference in 10 CFR 50.55a through the subject regulatory guides. Accordingly, the NRC staff developed RG 1.192, "Operation and Maintenance Code Case Acceptability, ASME OM Code," as well as RG 1.193, "ASME Code Cases Not Approved for Use."

The regulations at 10 CFR 50.55a(b)(6) incorporate by reference RG 1.192. Licensees may implement the Code Cases listed in RG 1.192 without obtaining further NRC review or approval if the Code Cases are used in their entirety with any supplemental conditions specified in the RG and the licensee's IST Code of Record is applicable to the Code Case. RG 1.193 lists Code Cases not approved for use.

If a licensee would like to use an ASME Code Case with an Edition or Addendum of the ASME Code to which it is not applicable, the licensee has the following options:

a. Have the alternative to use the Code Case, beyond its stated applicability, authorized by the NRC pursuant to 10 CFR 50.55a(a)(3), or

b. If the Code Case is applicable to an Edition or Addendum of the ASME Code later than the version of the Code being used by the licensee, the licensee could update to the later version of the Code pursuant to 10 CFR 50.55a(f)(4)(iv) or (g)(4)(iv) and then use the Code Case, provided the Code Case has been approved for use in the appropriate Regulatory Guide and incorporated by reference into 10 CFR 50.55a. Note that the later version of the ASME Code must also have been incorporated by reference into 10 CFR 50.55a, the licensee must update all related requirements of the respective Edition or Addenda, and the update must be specifically approved by the Commission.

The NRC may authorize the use of a Code Case that it has not yet been approved for use in RG 1.192 if a licensee requests the use of the code case under 10 CFR 50.55a(a)(3). The NRC may authorize the use of such a Code Case until a future revision to RG 1.192 accepts the use of the ASME Code Case. At that time, if the licensee intends to continue implementing the Code Case, they must follow all the provisions of the Code Case with the conditions specified in RG 1.192, if any. The authorization for a specific licensee to use a Code Case that is not listed in RG 1.192 does not authorize any other licensee to use the Code Case without submittal by the subsequent licensee of a request to implement an alternative to the ASME OM Code requirements under 10 CFR 50.55a(a)(3).

Code Cases OMN-1, "Alternative Rules for Preservice and Inservice Testing of Certain Electric Motor-Operated Valve Assemblies in Light-Water Reactor Power Plants," OMN-3, "Requirements for safety Significance Categorization of Components Using Risk Insights for Inservice Testing of LWR Power Plants," OMN-4, "Requirements for Risk Insights for Inservice Testing of Check Valves at LWR Power Plants," OMN-11, "Risk-Informed Testing for Motor-Operated Valves," and OMN-12, Alternate Requirements for Inservice Testing Using Risk Insights for Pneumatically and Hydraulically Operated Valve Assemblies in Light-Water Reactor Power Plants (OM-Code 1998, Subsection ISTC)," as accepted in RG 1.192 include risk-informed provisions that licensees may apply in IST programs. RG 1.175, "An Approach for Plant-Specific, Risk-Informed Decisionmaking: Inservice Testing," describes an acceptable alternative approach for applying risk insights from probabilistic risk assessment (PRA), in unction with established traditional engineering information, to make changes to a nuclear power plant's IST program. The approach described in RG 1.175 addresses the high-level safety principles specified in RG 1.174, "An Approach for Using Probabilistic Risk Assessment in Risk-Informed Decisions on Plant-Specific Changes to the Licensing Basis," and attempts to strike a balance between defining an acceptable process for developing risk-informed IST programs without being overly prescriptive. Until such time as a risk-informed regulation is promulgated and included in the regulations, the alternative approach described in RG 1.175 must be authorized by the NRC pursuant to 10 CFR 50.55a(a)(3)(i) on a plant-specific basis before being implemented by a given licensee. However, because 10 CFR 50.55a(a)(3)(i) places no restrictions on the scope of alternatives that the NRC may authorize, licensees may propose risk-informed alternatives to their entire IST program or may propose alternatives that are more limited in scope (e.g., for a particular system or group of systems, or for a particular group of components). However, with the issuance of RG 1.192, licensees may use specific risk-informed IST methods without first obtaining NRC staff review and approval. Section 8 discusses risk-informed IST in greater detail.

If RG 1.193 identifies a ASME Code Case as being unacceptable, the NRC is unlikely to approve a licensee request to use that specified Code Case (whether by exemption, approval of alternatives, or authorizing relief). Licensees requesting the NRC's approval to implement a Code case listed in the RG 1.193 must show, at minimum, that adequate protection to public health and safety is provided if the Code Case is applied by the licensee/applicant.

2.1.2 Conditions to the ASME OM Code

The NRC regulations incorporate by reference specific editions and addenda (up to the 2004 Edition including 2005 and 2006 addendas) to the ASME OM Code in 10 CFR 50.55a(b)(3), subject to the five conditions outlined below.

2.1.2.1 10 CFR 50.55a(b)(3)(i)—Quality Assurance

The OM Code references the use of either the ANSI/ASME NQA-1-1979, "Quality Assurance Program Requirements for Nuclear Facilities," or the owner's Appendix B, "Quality Assurance Criteria for Nuclear Power Plants and Fuel Reprocessing Plants," to 10 CFR Part 50 , "Domestic Licensing of Production and Utilization Facilities," as part of its individual provisions for a quality assurance program. However, ANSI/ASME NQA-1-1979 does not contain some of the quality assurance provisions and administrative controls governing operational phase activities that would be required in order to use ANSI/ASME NQA-1-1979 in lieu of an owner's 10 CFR Part 50, Appendix B quality assurance program description. The NRC originally endorsed ANSI/ASME NQA-1-1979 with the knowledge that it was not entirely adequate and that other commitments such as the ANSI/ASME standards must supplement it. Hence, ANSI/ASME NQA-1-1979 is not acceptable for use without the other quality assurance program provisions identified in TS and licensee quality assurance programs.

2.1.2.2 10 CFR 50.55a(b)(3)(ii)—Motor-Operated Valve Testing

This condition requires that licensees establish a program to ensure that motor-operated valves (MOVs) continue to be capable of performing their design-basis safety functions. The condition in 10 CFR 50.55a(b)(3)(ii) supplements the quarterly MOV stroke-time testing requirement in Subsection ISTC of the OM Code. Since 1989, the NRC has recognized that quarterly stroke-time testing is not sufficient to provide assurance of MOV capability under design-basis conditions. For example, in Generic Letter (GL) 89-10, "Safety-Related Motor-Operated Valve Testing and Surveillance," dated June 28, 1989, the NRC stated that stroke-time testing alone is not sufficient to provide assurance of MOV operability under design-basis conditions. Therefore, in GL 89-10, the NRC staff requested licensees to verify the design-basis capability of their safety-related MOVs and to establish long-term MOV programs. The NRC subsequently issued GL 96-05, "Periodic Verification of Design-Basis Capability of Safety-Related Power-Operated Valves," dated September 18, 1996, to provide updated guidance for establishing long-term MOV programs. The condition in 10 CFR 50.55a(b)(3)(ii) establishes a regulatory requirement for nuclear power plant licensees implementing the applicable editions and addenda to the ASME OM Code to establish programs to periodically assess the design-basis capability of MOVs within the scope of the IST program at nuclear power plants.

Code Case OMN-1, "Alternative Rules for Preservice and Inservice Testing of Certain Electric Motor-Operated Valve Assemblies in Light-Water Reactor Power Plants," for the ASME OM Code allows users to replace quarterly MOV stroke-time testing with a combination of MOV exercising at least every refueling outage and MOV diagnostic testing on a longer interval. In RG 1.192, the NRC addresses the acceptability of Code Case OMN-1 in lieu of the quarterly MOV stroke-time testing requirements in Subsection ISTC of the OM Code. The implementation of ASME OM Code Case OMN-1 as accepted in RG 1.192 can be used in satisfying the requirement in 10 CFR 50.55a(b)(3)(ii). ASME has incorporated ASME OM Code Cases OMN-1, and OMN-11, "Risk Informed Testing of Motor-Operated Valves," as Appendix III in the 2009 Edition to the ASME OM to replace quarterly MOV stroke-time testing with periodic exercising and diagnostic testing. In the future, the NRC staff will review the 2009 Edition of the ASME OM Code (including Appendix III) for incorporation by reference in 10 CFR 50.55a with any appropriate conditions. (Note: The details related to the ASME OM-2009 are for information only.)

2.1.2.3 10 CFR 50.55a(b)(3)(iv)(A), (B), (C), and (D)—Appendix II

This condition supplements the provisions in Appendix II, "Check Valve Condition Monitoring Program," to the OM Code. Subsection ISTC of the OM Code permits the use of Appendix II as an alternative to other testing or examination provisions of Subsection ISTC. If a licensee elects to use Appendix II, the provisions of Appendix II become mandatory in accordance with OM Code requirements. The conditions in 10 CFR 50.55a(b)(3)(iv) do not apply to the 2003 Addendum and later editions and addenda to the OM Code because the 2003 Addendum revised the earlier OM Code provisions on which this regulation was based to address the underlying issues that led the NRC to impose the condition.

The following conditions apply to ASME OM edition and addenda before 2003 addendum:

The condition in 10 CFR 50.55a(b)(3)(iv)(A) applies to the testing or examination of the check valve obturator movement to both the open and closed positions to assess its condition and confirm acceptable valve performance. The OM main committee approved the bidirectional testing of check valves for inclusion in the 1996 Addendum to the OM Code. The NRC agrees with the need for a required demonstration of the bidirectional exercising movement of the check valve disk. The single direction flow testing of check valves will not always detect degradation of the valve. The classic example of this faulty testing strategy is that separation of the disk would not be detected during forward flow tests. The separated disk could be lying in the valve bottom or another part of the system and could move to block flow or disable another valve. Appendix II did not require bidirectional testing of check valves in the 1996 through 2002 Addenda to the OM Code. Hence, the condition in 10 CFR 50.55a(b)(3)(iv)(A) was included so that an Appendix II condition monitoring program includes bidirectional testing of check valves to assess their condition and confirm acceptable valve performance (as is required by the OM Code).

The condition in 10 CFR 50.55a(b)(3)(iv)(B) applies to the length of the check valve test interval. Appendix II would permit a licensee to extend check valve test intervals without limit. A policy of prudent and safe interval extension dictates that any interval extension must be based on sufficient experience to justify the additional time. Condition monitoring and current experience may qualify some valves for an initial extension, whereas the trending and evaluation of the data may dictate reduction in the testing interval for some valves. Extensions of IST intervals must consider plant safety and be supported by the trending and evaluation of both generic and plant-specific performance data to ensure that the component is capable of performing its intended function over the entire IST interval. Thus, the condition in 10 CFR 50.55a(b)(3)(iv)(B) limits the time between the initial test or examination and the second test or examination to two fuel cycles or 3 years (whichever is longer), with any extension of this extension may not exceed one fuel cycle per extension with the maximum interval not to exceed 10 years. An extension or reduction in the interval between tests or examinations would have to be supported by trending and evaluation of performance data.

The condition in 10 CFR 50.55a(b)(3)(iv)(C) applies to a licensee who discontinues a condition monitoring program when using the 1995 Edition of the OM Code with the 1996 and 1997 Addenda. A licensee who discontinues the use of Appendix II is required to implement the requirements of Subsections ISTC 4.5.1 through ISTC 4.5.4 of the OM Code.

The condition in 10 CFR 50.55a(b)(3)(iv)(D) applies to a licensee who discontinues a condition monitoring program when using the 1998 Edition through the 2002 Addendum to the OM Code. A licensee who discontinues the use of Appendix II is required to implement the applicable provisions in Subsection ISTC.

2.1.2.4 10 CFR 50.55a(b)(3)(vi) —Exercise Interval for Manual Valves

This condition requires that manual valves must be exercised on a 2-year interval rather than the 5-year interval specified in paragraph ISTC-3540 of the 1999 through 2005 Addenda to the OM Code, provided that adverse conditions do not require more frequent testing. The 1998 Edition and earlier versions of the OM Code specified an exercise interval of 3 months for manual valves. The 1999 Addendum to the OM Code revised paragraph ISTC-3540 to extend the exercise frequency for manual valves to 5 years; however, the NRC staff did not agree that there was sufficient justification to extend the exercise interval for manual valves to 5 years. (See Federal Register Notice 67 FR 60520, 60531-32 (dated September 26, 2002). The condition in 10 CFR 50.55a(b)(3)(vi) does not apply to the 2006 Addendum to the OM Code because ASME revised the exercise interval in paragraph ISTC-3540 of the 2006 Addendum to the OM Code to 2 years for manually-operated valves.

2.1.3 Voluntary Use of Later Editions and Addenda to the ASME Code

10 CFR 50.55a(f)(4)(iv) states that IST programs for pumps and valves may meet the requirements set forth in subsequent editions and addenda to the OM Code that 10 CFR 50.55a(b)(3) incorporates by reference, subject to NRC approval. Licensees may use portions of editions or addenda provided that all related requirements of the respective editions or addenda are met.

When planning to use editions and addenda to the OM or Section XI Code that have not been incorporated by reference in the regulations, licensees must request authorization to use these later editions and addenda as an alternative to the regulations under 10 CFR 50.55a(a)(3).

The amount of written documentation needed for a request to use a later OM Code edition and addendum that 10 CFR 50.55a(b) incorporates by reference is significantly less than that necessary for other types of requests to use an alternative approach. For example, licensees are not required to provide specific justification for requests to use later OM Code editions and addenda that 10 CFR 50.55a(b) incorporates by reference. This is because the NRC has reviewed and accepted the provisions of those OM Code editions or addenda, with any appropriate modifications or limitations conditions, as part of the process for incorporation of the edition and addenda by reference in the regulations. If a licensee uses portions of a later OM Code edition and addendum, it must ensure that it meets all related requirements of the respective editions and addenda. The licensee should discuss the related requirements in its letter to the NRC. The regulations do not specify when the licensee should submit the letter, only that it should submit the letter before it uses the later OM Code edition and addendum. The staff issued Regulatory Issue Summary (RIS) 2004-12, "Clarification on Use of Later Editions and Addenda to the ASME OM Code and Section XI," dated July 28, 2004, and RIS 2004-16, "Use of Later Editions and addenda to the ASME Code Section XI for Repair/Replacement Activities," dated October 19, 2004, to clarify this matter.

2.1.4 Identification of Code Noncompliance

The attachment to RIS 2005-20, Revision 1, "Revision to NRC Inspection Manual Part 9900**, Technical Guidance, Operability Determinations & Functionality Assessments for Resolution of Degraded or Nonconforming Conditions Adverse to Quality or Safety," (ADAMS Accession No. ML073440103) dated April 16, 2008, includes guidance on resolving degraded and nonconforming conditions. Section 6.2 of the Inspection Manual (IM) Part 9900 includes guidance to a licensee that may discover a noncompliance with a regulation, such as noncompliance with ASME OM Code requirements. Noncompliance with regulations should be treated as a degraded or nonconforming condition, and the operability or functionality of the affected structures, systems or components (SSCs) should be assessed. If the noncompliance is not addressed by the operating license or the TS (i.e., the noncompliance has no impact on any TS function), the licensee should determine if the noncompliance raises an immediate safety issue.

Note: **Inspection Manual Part 9900 is being revised and will be reissued as NRC Inspection Manual, Manual Chapter (MC) 0326, "Operability Determination & Functionality Assessments for Conditions Adverse to Quality and Safety" (ADAMS Accession No. ML12346A480)

Common examples of ASME OM Code noncompliance that require the operability of the affected SSCs to be assessed result from failure to perform or meet an IST test required by a TS Surveillance Requirement (SR), or when components that perform TS functions fail or are discovered to be in a degraded condition when conducting an OM Code tests. In cases where the component's performance is nonconforming because the ASME Code required action range or limiting values are more conservative than the TS or safety analysis report (SAR) limits, the corrective action may be an analysis to demonstrate the specific nonconformance does not impair operability and the pump or valve will still perform its safety function. These actions would be accomplished in accordance with IM Part 9900 and the applicable edition and addenda of the OM Code.

In cases where a component does not meet the ASME OM Code and is therefore inoperable, because the component performs a TS function or a necessary and related support function per the TS definition of Operability, the component could be declared operable once the NRC authorizes the alternative test and the licensee has successfully completed the alternative test (if applicable). NRC authorization of an alternative test would not be retroactive because the NRC staff must authorize the alternative before it can be implemented.

2.1.5 ASME OM Code Interpretations

The ASME issues "Interpretations" to clarify provisions of the OM Code. Users submit requests for interpretation and, after appropriate committee deliberations and balloting, ASME issues response. Interpretations do not follow the same approval process as ASME OM Code and Code Cases. The Code interpretations provide the meaning or the intent of the existing requirements in the OM Code. Licensees should exercise caution when applying interpretations as they are not specifically part of the incorporation by reference into 10 CFR 50.55a and have not received NRC approval. The NRC recognizes that the ASME is the official interpreter of the Code, but the NRC will not accept ASME interpretations that, in the NRC's opinion, are contrary to NRC requirements or may adversely impact facility operations.

2.2 Scope of Inservice Testing Programs

General Design Criterion (GDC) 1, "Quality Standards and Records," (in Appendix A, "General Design Criteria for Nuclear Power Plants," to 10 CFR Part 50) and Criterion XI (in Appendix B, "Quality Assurance Criteria for Nuclear Power Plants and Fuel Reprocessing Plants," to 10 CFR Part 50) require that all components (such as pumps and valves) that are necessary for safe operation must be tested to demonstrate that they will perform satisfactorily in service. Among other things, GDC 1 requires that components that are important to safety must be tested to quality standards that are commensurate with the importance of the safety function(s) to be performed. Criterion XI requires, in part, that a test program shall be established to assure that all testing required to demonstrate that SSCs will perform satisfactorily in service is identified and performed in accordance with written test procedures.

In addition, 10 CFR 50.55a(f) requires that licensees must use the ASME OM Code for inservice testing of components that are covered by the Code. Each licensee has the responsibility to demonstrate the continued operability or functionality of all components within the scope of their IST program.

An IST program, including implementing procedures, is subject to the requirements of 10 CFR Part 50, Appendix B, and ASME OM Code Subsection ISTA. Changes to the scope, test methods, or acceptance criteria should be reviewed to the requirements of 10 CFR 50.59, 10 CFR 50.55a and 10 CFR 50.65, as appropriate.

The TS for some plants may include IST requirements that are more restrictive than the regulations.

2.2.1 Basis for Scope Requirements

The requirements for the scope of components to be included in an IST program are addressed in 10 CFR 50.55a(f). Specifically, 10 CFR 50.55a(f)(4) states, "Throughout the service life of a boiling- or pressurized-water-cooled nuclear power facility, pumps and valves which are classified as ASME Code Class 1, Class 2, and Class 3 must meet the inservice test requirements set forth in the ASME OM Code."

ASME Code Class 1 components include all components within the reactor coolant pressure boundary. RG 1.26, Revision 4, dated March 2007, provides guidelines for establishing the quality group classification (and ASME Code classification) for water-, steam-, and radioactive-waste-containing components of nuclear power plants, other than those in the reactor coolant pressure boundary (i.e., ASME Code Class 2 and Class 3 components).

The ASME OM Code is incorporated by reference in 10 CFR 50.55a(b)(3). The OM Code defines the scope by stating that IST programs shall include pumps and valves that are required to perform a specific function in (1) shutting down the reactor to a safe shutdown condition, (2) maintaining the safe shutdown condition, or (3) mitigating the consequences of an accident. The scope of the OM Code also covers pressure relief devices that are used to protect systems (or portions of systems) that perform a required safety-related function. Therefore, the scope of components to be included in an IST program must encompass ASME Code Class 1, 2, and 3 components that are covered in Subsection ISTA of the ASME OM Code.

Subsection ISTA-1100 of the OM Code refers to components that are "needed to mitigate the consequences of an accident." This statement is intended to provide confidence that the health and safety of the public will be protected in the event of certain accidents and anticipated transients at a nuclear power plant. The term "accident" is also used throughout the Commission's regulations. For example, Appendix B to 10 CFR Part 50 establishes quality assurance requirements for the design, construction, and operation of "structures, systems, and components that prevent or mitigate the consequences of postulated accidents that could cause undue risk to the health and safety of the public." Similarly, 10 CFR Part 100, "Reactor Site Criteria," describes structures, systems, and components that must be designed to remain functional during and following a "safe shutdown earthquake" as those necessary to ensure (1) the integrity of the reactor coolant pressure boundary, (2) the capability to shut down the reactor and maintain it in a safe shutdown condition, or (3) the capability to prevent or mitigate the consequences of accidents that could result in potential offsite exposures comparable to the guideline exposures.

In establishing such requirements, the Commission uses the term "accident" to describe a broad range of possible adverse events at a nuclear power plant. Therefore, although most of the accidents of concern to IST are addressed in the accident analyses chapter, licensees should be aware that the plant's final safety analysis report (FSAR) may address other accident analyses that need to be considered within the context of IST.

Thus, an introductory section of the IST program document submitted to the NRC for each existing plant should state the plant's safe-shutdown condition (e.g., hot standby, hot shutdown, cold shutdown).

Components within the scope of 10 CFR 50.55a are included in the scope of 10 CFR 50.65, "Requirements for Monitoring the Effectiveness of Maintenance at Nuclear Power Plants" (the Maintenance Rule). Licensees may elect to consolidate testing for pumps and valves, designating any non-Code components as such in the IST program.

The plant's FSAR defines the equipment that is necessary to meet specific functions. If the FSAR indicates that a system or component is Code Class 1, 2, or 3, that system or component is within the scope of 10 CFR 50.55a. By contrast, if the FSAR states that a system or component is designed, fabricated, and maintained as Code class at the option of the Owner as permitted by Subsection ISTA-1320, the application of the related OM Code requirements is also optional.

Tables 2.1 and 2.2 (which appear at the end of this chapter) provide examples of systems and components that licensees typically include in their IST programs. These tables are not intended to be all-inclusive, but they may form the basis for the initial review of a licensee's IST program scope.

Figure 2.1, "Flow Chart- Development of Inservice Testing Program for Pumps and Valves," provides a quick reference to regulatory requirements for development of the IST program for pumps and valves. For complete details, see 10 CFR 50.55a.

2.2.2 Examples of Omitted Components

During IST program reviews and inspections, the staff has noted that licensees do not always include the necessary equipment in the scope of their IST programs. Licensees should review their IST programs to ensure adequate scope. Components that are frequently erroneously omitted from IST programs include the following examples:

(a)   BWR scram system valves
(b)   control room chilled-water system pumps and valves
(c)   accumulator vent valves or motor-operated isolation valves
(d)   auxiliary pressurizer spray system valves
(e)   boric acid transfer pumps
(f)   valves in the emergency boration flow path
(g)   control valves that have a required fail-safe position
(h)   valves in mini-flow lines
(i)   control rod drive (CRD) system check valves
(j)   keep fill systems
(k)   excess flow check valves

Licensees should review the safety significance of these components to ensure that their IST is adequate to demonstrate their continued operability. Licensees should also recognize that the pumps and valves listed above do not apply to every plant and do not satisfy the scope required by Subsection ISTA for all plants. For example, items c, d, e, and f do not apply to BWRs. Each licensee should review the list and determine which items apply to its facility.

2.2.3   Testing of Non-Code Components

As discussed above, licensees are required to test safety-related components to demonstrate that they will perform satisfactorily in service in accordance with 10 CFR Part 50, Appendices A and B. The IST program for components within the scope of the ASME Code is addressed in 10 CFR 50.55a.

An IST program is also a reasonable vehicle to periodically demonstrate the operational readiness of pumps and valves that are not covered by the Code, but are within the scope of 10 CFR Part 50, Appendices A and B. Thus, if a licensee voluntarily chooses to include non-Code components in its ASME Code IST program (or some other licensee-developed testing program) and, as a result, is unable to meet certain Code provisions, the regulations (10 CFR 50.55a) do not require the licensee to submit a relief request to the NRC.

Nonetheless, the licensee should maintain documentation that provides assurance of the continued operational readiness, or as required the continued functionality of the non-Code components through the performed tests. Such documentation should be available for staff inspection at the plant site.

For example, the emergency diesel generator air start system is typically not within the scope of the Code. However, emergency diesel generator air start, cooling water, and fuel oil transfer systems are considered safety-related and, as such, Appendices A and B to 10 CFR Part 50 require that they must be included in the scope of a component testing program and must undergo the required testing. Licensees may implement deviations from the Code for non-Code components without NRC review and approval. A notation in the licensee's IST program document would help to identify the deviations and clarify that they relate to non-Code components.

2.2.4   Commitments to Include Components in IST Programs

The licensee is responsible for determining whether a component requires to be included within the IST program, or whether that classification is optional under Subsection ISTA-1320. Specifically, Subsection ISTA-1320 states that optional construction of a component within a system boundary to a classification higher than the minimum class established in the component design specification shall not affect the overall system classification by which applicable rules are determined. Thus, if a licensee changes the code classification pursuant to 10 CFR 50.59, the pumps and valves may remain as "augmented components" (denoted as non-Code) in the IST program. (Note that NRC approval of a licensee amendment may be necessary, as determined by the evaluation conducted in accordance with 10 CFR 50.59.) Regulatory Guide 1.187, "Guidance for Implementation of 10 CFR 50.59, Changes, Tests, and Experiments," (ADAMS Accession No. ML003759710) provides guidance for 10 CFR 50.59 implementation.

2.3   Code Class Systems Containing Safety-Related Pumps and Valves

The plant safety analysis report (SAR), TS, and other documents list the systems and components that are necessary to function to support the safe operation and shutdown of the plant. Tables 2.1 and 2.2 (which appear at the end of this chapter) list systems and components typically included in the IST programs for PWRs and BWRs. These tables are not intended to apply to all plants. The listed systems and components are not considered safety-related at every plant, and are not necessarily classified as Code Class 1, 2, or 3. For information on quality group and Code classifications, see RG 1.26 and Section 3.9.6 of NUREG-0800, "Standard Review Plan." The licensee's safety analysis generally contains a section describing the Code classification of components. The IST program scope should be developed to be consistent with the SAR.

2.4   IST Program Document

Within this discussion of the IST program document, Section 2.4.1 applies to pumps, while Section 2.4.2 applies to valves. These sections describe the information that licensees generally need to prepare the related sections of the IST program document.

The OM Code includes the rules for inservice testing of nuclear power plant components:

- Subsection ISTA includes general requirements for testing components.

- Subsection ISTA-3200 states that the owner shall file IST plans with the regulatory authorities having jurisdiction at the plant site.

- Subsection ISTA-9000 addresses the records and reports that are required for these inspection and testing programs.

- Subsection ISTA-9210 states that the owner shall prepare plans for preservice, and inservice examinations and tests to meet the requirements of the OM Code.

- Subsection ISTA-9220 states that licensees shall prepare examination, test, replacement, and repair records in accordance with the requirements of the respective articles of the OM Code.

- Subsections ISTB-9000 and ISTC-9000 include additional guidance for the information that the IST program document must include for pumps and valves that perform safety functions.

- Nonmandatory Appendix A, "Preparation of Test Plans," and Supplement to Appendix A to the ASME OM Code gives voluntary guidance for licensees to use in preparing their inspection and test plans.

Licensees have found that pump and valve tables are a convenient format for the information. These tables typically include sufficient information to allow NRC inspectors to determine whether the testing complies with the Code requirements for test method and frequency. The tables also could note applicable NRC positions or recommendations for each pump or valve.

The NRC intends that the IST program should reflect design modifications and other activities performed under 10 CFR 50.59 that relate to pumps and valves within the scope of the program. Thus, the staff recommends that the program plan submitted to the NRC should include documentation of the use of positions contained herein, and Code Cases.

2.4.1 Pumps

In preparing pump tables, licensees should consider the following information, which includes headings and a description of the text that licensees could include under each heading.

Title: List the applicable plant and unit.

Page number: Include the page number and total number of pages in the program document or the relevant section, such as "Page 15 of 135."

Program revision or revision date: List the program or page revision number and date (on each page). List the revision number for each program change submitted.

System, Code class, and group: List the plant system, Code class, and pump group, and briefly describe the service of the pump.

Pump identification: List a unique identifier for each pump. This identifier should be used consistently in all IST program documentation and design information such as system piping and instrument diagrams (P&IDs), test procedures, and relief requests.

Piping and instrument diagram (P&ID) number: List the applicable P&ID or figure that depicts the pump in the system.

Drawing coordinates: List the coordinates of the pump on the P&ID.

Test parameters: List each of the five parameters in Tables ISTB-5121-1, ISTB-5221-1, ISTB-5321-1, and ISTB-5321-2 for each pump. A column or a footnote is typically used to list factors that affect testing. List a relief request or Code Case number where the testing will not be performed in accordance with the Code. Notes can be used where Code testing would otherwise be required. A relief request is not required if the test requirement is exempted by the Code.

Relief request(s): List any applicable relief requests in the pump table. Table 2.3 provides an example of a data table for pumps.

2.4.2 Valves

In preparing valve tables, licensees should consider the following information, which includes headings and a description of the text that licensees could include under each heading. Table 2.4 lists common abbreviations used in valve data tables.

Title: List the applicable plant and unit.

Page number: Include the page number and total number of pages in the program document or the relevant section, such as "Page 15 of 135."

Program revision or revision date: List the program or page revision number and date (on each page). List the revision number for each program change submitted.

System, Code class, and group: List the plant system, Code class, and valve group, and briefly describe the service of the valve.

Valve identification: List a unique identifier for each valve. This identifier should be used consistently in all IST program documentation and design information such as P&IDs, test procedures, and relief requests. If valves such as excess flow check valves are grouped together in the table, the number of valves and the valve number must be clearly indicated.

P&ID number: List the applicable P&ID or figure that depicts the valve in the system.

Drawing coordinates: List the coordinates of the valve on the P&ID.

Valve type: List the valve type (i.e., gate, globe, check, relief).

Valve size: Specify the valve size in inches, fractions of an inch, or metric units.

Actuator type: List the type of valve actuator (i.e., motor, solenoid, pneumatic, hydraulic, self) with the type and function of each valve.

Code category: Specify the Code category (or categories), as defined in Subsection ISTC-1300. This determines the applicable subsections of the Code. For example, a motor-operated gate valve could be in Code Category A or B, while a self-actuated check valve could be in Category C or A/C.

Active/Passive: State whether a valve is active or passive, as defined in Subsection ISTC-2000. Requirements vary based on the function of the valve. A valve need not be considered active if it is only temporarily removed from service or from its safety position, such as manually opening a sample valve for a short time to take a sample, while maintaining administrative control over the valve. If the plant is in an operating mode that does not require a passive valve to be maintained in its "passive" (safety) position, the position of the valve may be changed without imposing IST requirements on the valve. By contrast, if a valve is routinely repositioned during power operations (or has an active safety function), it is an active valve. If a valve is repositioned to create a new alignment (e.g., as a corrective action for a condition of another valve in the line), an evaluation (considering the impact on the IST program) may be required to ensure operational readiness before positioning the valve in a new position, as determined on a case-by-case basis.

Safety position: List the safety function position(s), and specifying both positions for valves that perform a safety function in both the open and closed positions. Valves must be exercised to the position(s) required to fulfill their safety function(s). Check valve tests must include both open and close tests.

Tests performed: Specify which tests are to be performed on each valve. Test frequency: List the actual frequency for each test to be performed. If it would be impractical or burdensome to perform the test at the frequency specified in the Code, reference cold shutdown or refueling outage justifications or relief requests for the alternative test frequency.

Relief requests and cold shutdown/refueling outage justifications: List any applicable relief request(s). In addition, when the testing is deferred to cold shutdowns or refueling outages, reference the technical justification (cold shutdown justification or refueling outage justification) for the test frequency.

Remarks: Include any pertinent information that is not stated elsewhere in the table such as a brief functional description of the valve.

2.4.3 Piping and Instrument Diagrams

The staff recommends that licensees' program submittals should include P&IDs or system drawings to assist in locating the pumps and valves that are included in the program and such drawings should be the latest revision at the time the program is submitted to the NRC. This information will assist the staff in reviewing relief requests or proposed alternatives. Inservice inspection boundary system drawings and isometrics, or reduced-size drawings, are suitable for inclusion in the program document. If the reduced-size drawings are not complete P&IDs, the staff may request a set of full-size drawings for use in evaluating relief requests. A partial submittal of the program containing relief requests could include applicable drawings to support the relief requests or to supersede previous IST program drawings. Licensees need not update their program drawings regularly, but if drawings change because of modifications, or if the changes affect relief requests, the staff recommends that licensees should revise and resubmit the drawings in the next periodic submittal of revisions to the program document. The staff also recommends that licensees should include applicable drawings with relief request submittals that are very detailed and are submitted to supplement the IST program. Such drawings are helpful because the NRC's technical staff who review relief requests do not maintain a set of SARs for each plant and do not receive a copy of the IST program plan (which generally contains the applicable drawings). Drawings are helpful in reviewing relief requests, regardless of whether they are submitted as part of the program document or as an attachment applicable to any relief request or proposed alternative.

2.4.4 Bases Document

The NRC staff recommends that each licensee create a bases document for the IST program. A paper discussing the creation and management of a bases document is included in Supplement 1 to NUREG/CP-0123, "Proceedings of the Second NRC/ASME Symposium on Pump and Valve Testing," dated November 1992. Bases documents have typically included a description of the methods used in preparing the IST program, with a list of each pump and valve in a system within the boundaries for a Code class, the basis for including (or excluding) the pump or valve, and the basis for the testing applied to each component.

Although not required by the NRC, the bases document may help licensees ensure the continuity of their IST programs when the responsibilities of personnel or groups change. A bases document will also enable the plant staff to clearly understand the reasons that the components are either in the program or not, as well as the basis for testing (or not testing) certain functions. Although not a "licensing-basis document" (unless the licensee takes action to make the document part of the licensing basis for a plant) the bases document is useful reference for licensee reviews performed under 10 CFR 50.59 when changes are being considered for a facility.

2.4.5 Deferring Valve Testing to Cold Shutdown or Refueling Outages

Exercising valves on a cold shutdown or refueling outage frequency does not constitute a deviation from the Code. Subsection ISTC-3520 provides guidance for testing valves during cold shutdown or refueling outages if it is impractical to test during operation. The licensee should list the affected valves in the program document and include cold shutdown or refueling outage justifications for each affected valve or group of valves. The staff recommends that licensees should include these cold shutdown and refueling outage justifications in their IST program submittals to the NRC.

Check valves that can be stroked quarterly, but must be monitored by a nonintrusive technique to verify full stroke, may be full-stroke tested during cold shutdown or refueling outages if another method of verifying full-stroke exists during such plant conditions. The NRC would not require a licensee to invest in nonintrusive equipment for the purpose of testing check valves quarterly (instead of testing them during cold shutdown or refueling outages), even though the use of nonintrusive techniques is recommended where practical.

A licensee may request relief from quarterly testing where such testing would impose a hardship (e.g., entering a limiting condition for operation of 3 to 4 hours in duration, repositioning a breaker from "off" to "on," and necessitating for manual operator actions to restore the system if an accident were to occur while the test was in progress). For such situations, the risk associated with quarterly testing may outweigh the benefits that might otherwise be achieved. (Section 3.1.2 gives guidance on these types of situations.) Thus, it is appropriate for licensees to weigh the safety impact against the benefits of testing as a basis for deferring testing from a quarterly frequency to cold shutdown or refueling outages. NUREG/CR-5775, "Quantitative Evaluation of Surveillance Test Intervals Including Test-Caused Risks," dated March 1992, (ADAMS Accession No. ML027410457) describes a method for making this comparison.

In the event of a planned or unplanned maintenance outage, a licensee may decide to test some or all valves in a cold shutdown mode, rather than waiting for the refueling outage. In making this decision, the licensee should consider the duration of the shutdown and the extent of other outage activities. The requirements of Subsection ISTC-3560 for testing valves in systems that are out of service may apply for extended outages that last for several months. Guidance on minimizing shutdown risk also may apply for extended outages.

Impractical conditions justifying test deferrals may include the following situations that could result in an unnecessary plant shutdown, cause unnecessary challenges to safety systems, place undue stress on components, cause unnecessary cycling of equipment, or unnecessarily reduce the life expectancy of the plant systems and components:

- inaccessibility
- testing that would require major plant or hardware modifications
- testing that has a high potential to cause a reactor trip
- testing that could cause system or component damage
- testing that could create excessive plant personnel hazards
- existing technology that will not give meaningful results

In the licensing process, the NRC staff weighs the possible safety consequences and benefits of performing a required test as part of TS surveillance, including circumstances in which one train is out of service. Nonetheless, any related guidance provided by the staff does not supersede the TS requirements. For example, if testing is specified as part of the TS surveillance, the cycling of nonredundant valves in a remaining operable train may not be deferred to the next cold shutdown when one train is out of service, even though their failure would cause a loss of total system function. In this case, a TS change or enforcement discretion would be necessary to defer testing.

The NRC expects licensees to comply with required IST test frequencies. The Code does not require documentation for valves that are not tested during a cold shutdown outage other than as required for maintaining the IST schedule. The NRC does not have a position on the efforts a licensee expends in performing cold shutdown valve testing during a short outage. The staff, however, expects licensees to expend a reasonable "good faith" effort.

This issue is further discussed in Sections 3 and 4, which give guidance on deferring testing.

2.5   Relief Requests and Proposed Alternatives

Alternative Requests

Licensees can request that the NRC authorize an alternative to an OM or Section XI Code requirement in accordance with 10 CFR 50.55a(a)(3). Requests made under 10 CFR 50.55a(a)(3) are more specifically called "alternatives."

The OM Code establishes the requirements for preservice testing and IST and the examination of certain components to assess their operational readiness in light-water reactor nuclear power plants. These requirements apply to pumps, valves, pressure relief devices, and snubbers within the scope of the OM Code. The requirements are constantly being reviewed and improved in order to meet the basic function of maintaining the safe and reliable operation and maintenance of nuclear power plants.

It is understood that not all plants are designed the same. It is also understood that the general requirements developed in the OM Code may not be applicable or that complying with these requirements may be difficult. Licensees may use proposed alternatives to the OM Code provided that (1) the alternative would provide an acceptable level of quality and safety under 10 CFR 50.55a(a)(3)(i); or (2) compliance with the specified requirements would result in hardship or unusual difficulty without a compensating increase in the level of quality and safety under 10 CFR 50.55a(a)(3)(ii). Hardships generally involve reductions in radiation exposure to as low as reasonably achievable, challenges to operators or plant equipment, components that are somewhat unique in design such as jockey (waterleg) pumps, or systems where pump flow is fixed and cannot be adjusted.

Licensees must not implement proposed alternatives to the OM Code requirements under 10 CFR 50.55a(a)(3) until the NRC staff completes its evaluation. For example, if a licensee proposes to implement a pump vibration program based on the use of spectral analysis rather than the OM Code-specified method, the licensee must continue to meet the OM Code requirements until the NRC staff completes its evaluation.

Relief Requests

Licensees can request that the NRC grant relief from an OM or Section XI Code requirement in accordance with 10 CFR 50.55a(f)(5)(iii), 10 CFR 50.55a(f)(5)(iv), 10 CFR 50.55a(g)(5)(iii), and 10 CFR 50.55a(g)(5)(iv). Requests made under 10 CFR 50.55a(f)(5) and 10 CFR 50.55a(g)(5) are called "relief."

The regulation at 10 CFR 50.55a(f)(4) requires licensees to test pumps and valves in the IST program to the "extent practical" within the limitations of the design, geometry, and materials of construction. The regulations at 10 CFR 5.55a(f)(5)(iii)–(iv) and 10 CFR 50.55a(f)(6)(i) use the term "impractical" instead of "extent practical." The terms "extent practical" and "impractical" apply to test requirements in the OM Code that licensees cannot perform because of the design, geometry, and materials of construction of the pump or valve. For example, ASME OM Code, Subsection ISTC-5131, "Valve Stroke Testing," requires that the limiting stroke time for power operated valves be specified by the licensee and measured within limits based on the full-stroke time of the valves. At some plants, the scram discharge volume vent and drain valves are not designed to be individually actuated. These valves are required by TS to close within a specified time (45 seconds for some plants) upon receipt of a scram signal. The valves are tested quarterly by cycling the valves to ensure operability and performing a valve sequence response time test during each refueling outage. This testing is essentially a design basis test of the valve combination. Requiring these valves to be stroke timed individually is impractical and places a burden on the licensee because of the extensive modification that would be required to the system to individually stroke the valve. In addition, jumpering the control circuit during plant operation to test these valves individually would be impractical because of the potential for a reactor scram. Some licensees may have difficulty fully implementing these ISTC-5131 required tests, and, in certain cases, because of the impracticality of implementation, a request for relief under 10 CFR 50.55a(f)(5) would be appropriate.

In accordance with the regulations, when updating a program to a later edition of the OM Code, licensees must implement the updated program at the beginning of a 120-month interval. The regulations state that in cases in which a licensee determines that an OM Code-specified pump or valve test is impractical and is not included in the revised IST program, it must submit a relief request demonstrating the basis for its determination to the NRC no later than 12 months after the previous 120-month interval ends, or 12 months after the current interval starts. However, experience has shown that licensees also identify impractical test provisions throughout the interval. In such cases, licensees may request relief as soon as they identify the condition.

Because the requirements are impractical, the licensee would test the applicable components using the method proposed in the relief request in the period of time from the beginning of the new interval (or from the time of identification) until the NRC staff completes its evaluation.

2.5.1 Justifications for Relief or Alternatives

In determining whether to grant relief from the Code requirements or to authorize alternatives, the NRC staff considers the merits of the submitted technical information. In requesting relief or use of an alternative, the licensee would typically identify the specific Code requirement and associated paragraph for which relief or use of an alternative is requested, describe the proposed alternative(s), describe the basis for relief or authorization of the proposed alternative(s), and clarify the burden that would result if the NRC enforced the specified requirements. Situations that warrant granting relief or authorizing alternatives (as determined by the staff in previous safety evaluations for plant-specific requests) may include the following examples:

(1)   In complying with the Code requirements, the licensee would not obtain information that would be more useful than the information that is currently available. For example, installing an analog gauge with a range of three times the reference value (or less) to comply with Code requirements may not yield more accurate readings than those provided by the gauge that is presently installed (see Section 5.5.1).

(2)   Compliance with the Code is impractical because of design limitations. Imposition of the Code requirements would require significant system redesign and modifications. For example, a flow meter does not meet the accuracy requirements of ISTB-3510 and Table ISTB 3510-1 because the present system configuration does not have a straight section of pipe of sufficient length in which to measure flow accurately (see Section 5.5).

(3)   The required measurements or appropriate observations cannot be made because of physical constraints. Examples include a component located in an area that is inaccessible during power operation or a pump that is totally immersed in system fluid.

(4)   The need to keep personnel radiation exposure as low as reasonably achievable (ALARA) may present an adequate justification. The licensee should include information about the general area radiation field, local hot spots, plant radiation limits and stay times, the amount of exposure personnel would receive in doing the testing, and the safety significance of deferring testing or performing an alternative method. ALARA relates to controlling exposure during an activity, not specifically to eliminating activities; however, it may be a basis for relief or for deferring a test on the basis of hardship when exposure limits are prohibitive for performing testing (or possibly for accessing a valve for repair in the event that it could fail during a test). If the exposure limits are prohibitive, the licensee should defer testing to cold shutdown or refueling outages during which the exposure limits would no longer be prohibitive. ALARA is part of an overall program, including activities such as IST, as required by 10 CFR 20.1101, "Radiation Protection Programs." The NRC has not established "predetermined acceptable limits" for deferring an IST activity, based on maintaining occupational exposure ALARA.

(5)   Testing as required by the Code could cause significant equipment damage. For example, shutting off cooling flow to an operating pump by exercising a valve in the cooling flow path could damage the pump.

(6)   Failure of a component during testing could disable multiple trains of a reactor safety system. For example, a motor-operated suction valve common to both trains of high-pressure safety injection could not be tested during power operation because a failure of the valve would result in both trains being out of service.

Inconvenience or administrative burden is not, in and of themselves, adequate justification for deviating from the Code requirements. Similarly, entering a TS limiting condition for operation (LCO) is not, in and of itself, adequate justification for deviating from the Code-specified frequency, except when entering the LCO would be prohibited because the total system function would be out of service.

2.5.2   Categories of Relief or Alternative Requests

The NRC staff categorizes relief or alternative requests as follows:

- General: A general relief or alternative request is appropriate when the requested relief or alternative applies to a broad range of similar components in the program, such as all pumps or all containment isolation valves.

- Specific: A relief or alternative request is specific when the requested relief or alternative applies only to a single component or a specified group of similar components in the program, such as service water pump discharge check valves.

2.5.3   Content and Format of Relief or Alternative Requests

As a minimum, the staff recommends that each relief or alternative request should include the following information:

- Title and relief or alternative request number: Licensees should title each relief or alternative request and specify a unique identifier. The identifier should remain unique to avoid confusion with later revisions. Examples include (1) "Relief Request Number 1," (2) "Safety Injection Pumps Relief Request," or (3) "Check Valves in Series Relief Request."

- Page number: List the page number and total number of pages in the program document or the relevant section, such as "Page 15 of 135."

- Program revision or page revision date: List the program or page revision number and date (on each page). List the revision number for each program change submitted.

- System and Code class: List the plant system and Code class of the system in which the component is located.

- Pump/valve category or group: List the ASME category or group for each pump or valve (i.e., A, A/B, A/C, B, C, or D).

- Component identification: List the identification number for each component in a specific relief or alternative request. Each individual component need not be listed in a general relief or alternative request, such as one for all pumps in the IST program. However, the staff recommends that the list of program components (pump or valve table) should include the relief or alternative request number.

- Component function: Briefly describe the functions of the affected components and specify the function that is the subject of the relief or alternative request.

- ASME Code test requirement(s): List and describe the Code requirement(s) from which relief or alternative is being requested.

- Basis for relief or alternative: Clearly state the legal basis under which relief or an alternative is requested, and then explain the reasons why complying with the Code requirements is impractical, poses a hardship, or otherwise should not be enforced. Include all information that the NRC staff might need to complete its review. For example, most relief requests for check valves list the test direction(s) for which relief is requested.

- Proposed alternative testing: Clearly and thoroughly discuss the proposed alternative in sufficient detail to clearly demonstrate why it is a reasonable alternative to the Code requirement, and provide a technical basis for its acceptability.

- Drawings and/or diagrams: If the relief request or alternative testing is complex, or if drawings or diagrams are available for further clarification, include them in the relief or alternative request, or include them in the IST program document and reference them in the relief or alternative request.

- References: List references to SAR sections, technical specifications, and other pertinent documents. Any document referenced in the relief or alternative request should be submitted to the NRC on the plant docket. If a document is not docketed but contains pertinent information, the relief request should explicitly include the information (if it is not readily available to the staff and the public), rather than merely referencing the document.

To improve the effectiveness and efficiency of the request process, the Nuclear Energy Institute (NEI) developed a white paper entitled, "Standard Format for Requests from Commercial Reactor Licensees Pursuant to 10 CFR 50.55a, Revision 1," dated June 7, 2004 (ADAMS Accession No. ML070100400). This white paper provides useful guidance for determining the appropriate regulatory requirement under which a "relief request" is submitted for NRC approval, as well as the appropriate format and content to use in the request. The term "relief request" is used loosely in this instance to denote the various types of submittals allowed by

10 CFR 50.55a, including alternatives to the regulation [10 CFR 50.55a(a)(3)], impractical relief requests [10 CFR 50.55a(f)(6)(i)], and requests to use later Code Editions and Addenda [10 CFR 50.55a(f)(4)(iv)]. The NRC staff has reviewed the NEI White Paper and encourages licensees to use the specified format and content.

2.5.4 Revising NRC-Authorized Relief or Alternative

RG 1.187, "Guidance for Implementation of 10 CFR 50.59, Changes, Tests, and Experiments," dated November 2000, provides guidance related to use of 10 CFR 50.59 process. This 10 CFR 50.59 process does not allow the licensee to change an NRC granted or authorized relief request or alternative. The NRC must authorize any change to an NRC-authorized 10 CFR 50.55a(a)(3) alternative unless the requirements of the ASME Code can be met. For example, many licensees created a technical requirement manual (TRM) to control certain provisions relocated from TSs. Licensees relocated snubber examination and testing requirements from the TS to the TRM. The TRM requirements are controlled using the criteria in 10 CFR 50.59. The regulations at 10 CFR 50.59 require licensees to evaluate proposed changes to their facilities for the effects of these changes on the licensing basis of the plant, as described in the final safety analysis report (FSAR) (as updated) and to obtain prior NRC approval for changes that meet specified criteria as having a potential impact upon the basis for the issuance of the operating license. In the case of snubber examination and testing, the NRC has authorized the use of the TRM snubber examination and testing requirements in lieu of the ASME code requirements at numerous operating plants. The NRC authorized the use of the requirements contained in the TRM as an alternative to the ASME code requirements. The use of an alternative as authorized by the NRC becomes a regulatory requirement; therefore, the NRC staff must review and approve changes to these requirements under 10 CFR 50.55a(a)(3).

Nuclear Energy Institute (NEI) 96-07, "Guidelines for 10 CFR 50.59 Implementation," Revision 1, dated November 2000, states that licensees' activities that are controlled by the regulations at 10 CFR 50.55a take precedence over 10 CFR 50.59. RG 1.187 endorses NEI 96-07, Revision 1. Similarly, Section D, "Implementation," of RG 1.187 states that 10 CFR 50.59 cannot be used in those cases in which a licensee proposes an acceptable alternative method for complying with the specified portion of the NRC's regulation.

2.5.5 NRC Temporary Verbal Authorization of an Alternative Request

On rare occasions, the NRC may grant verbal authorizations as an alternative under 10 CFR 50.55a(a)(3) when, because of unforeseen circumstances, licensees need NRC authorization before the agency is able to issue its written safety evaluation as described in NRC document LIC-102, "Relief Request Reviews," (ADAMs Accession No. ML09138059). Temporary verbal authorization for an alternative under 10 CFR 50.55a(a)(3) is subject to the following:

- The proposed alternative is in writing, and all the information that the NRC requires to complete the safety evaluation has been docketed.

- An identified need for the verbal authorization is recognized given the circumstances of the licensee's request.

- The NRC has completed its review and determined that the proposed alternative is technically justified, but the agency has not yet formally documented it in a safety evaluation.

- The technical branch and reactor licensing branch chiefs have agreed to the verbal authorization.

Verbal authorization is most likely conveyed in a telephone conversation. As such, appropriate NRC personnel who are normally involved in authorizing the alternative must be present in the telephone conversation. The NRC project manager should promptly (i.e., in 1 or 2 days) generate a record of the conversation; this record will meet the definition of an Official Agency Record (OAR) and must be entered into the ADAMS and made publicly available. The NRC should issue the final written authorization within 150 days after giving verbal authorization.

2.5.6 NRC Approval of Proposed Alternative Similar to Prior NRC-Approved Alternative

Licensees occasionally submit alternative requests that are very similar to NRC-authorized alternative requests for the previous 10-year IST intervals when updating their IST program in accordance with 10 CFR 50.55a(f)(4)(ii). This practice is acceptable provided that the licensee compares the requirements between the old and new OM Codes and evaluates whether changes to the alternative request are necessary. For example, the OM Code has new provisions added for exercising check valves such as disassembly and condition monitoring programs. Addressing the check valve disassembly and condition monitoring programs in the alternative request may be appropriate if these provisions were not included in the OM Code upon which the original alternative request was based. Furthermore, the addition of disassembly and condition monitoring programs to the OM Code may eliminate the need for the alternative request.

Licensees also should review new code cases before submitting an alternative request for updated IST programs. For example, Code Case OMN-9, "Use of Pump Curve for Testing," provides an alternative method for testing centrifugal and vertical line shaft pumps when the licensee is unable to obtain a specific reference value in accordance with Subsection ISTB of the OM Code. The NRC conditionally approved Code Case OMN-9 in RG 1.192, Revision 0. Code Case OMN-16, "Use of Pump Curve for Testing," incorporates all the conditions specified for approval of Code Case OMN-9 and published in the 2006 addendum of the ASME OM Code. The use of Code Case OMN-9 or OMN-16 may eliminate the need for an alternative request. As of the issuance date of this NUREG, Code Case OMN-16 has not been accepted in the revision to RG 1.192 incorporated by reference in 10 CFR 50.55a.

2.6 IST Program Documents

The applicable ASME OM Code, Subsection ISTA, General Requirements, provides documentation requirements such as the following:

- ISTA-3200(a) requires that IST of pumps and valves plans shall be filed with the regulatory authorities having jurisdiction at the plant site.

- ISTA-9000, "Records and Reports," provides the requirements for preparation, submittal, and retention of records and reports.

- Nonmandatory Appendix A and Supplement to Nonmandatory Appendix A describe voluntary guidance for licensees for development of IST of pumps and valves plans.

As long as the IST program is consistent with the regulations, ASME Code relief is not required. That is, deletions from or additions to the IST program do not necessarily require NRC approval. The burden is on each licensee to verify that its IST program is complete and includes all components that require IST, and that all such components are tested to the extent practical. If a licensee deletes a particular component from its IST program, the staff recommends that the licensee should document the reason in an appropriate place.

The staff expects each licensee to maintain its IST program up-to-date and ensure that it remains consistent with changes in plant configuration. If a particular relief request is no longer required because of changes in hardware, system design, or new technology, the licensee is expected to revise its program to withdraw the relief request. Conversely, if a system modification results in the addition of a component to the IST program, the licensee should ensure that it meets the Code requirements, or that a relief request is submitted for NRC review and approval, as appropriate.

Licensees not meeting ISTA-3200(a) must submit appropriate documents containing pumps and valves IST plans and submit a request for relief to the NRC pursuant to 10 CFR 50.55a(a)(3). IST program documents submitted to the NRC are used to prepare for IST inspections and to address other licensing actions that may arise. Between a licensee's 10-year interval program submittals, the NRC would like to receive up-to-date program documents when the licensee makes significant changes to the IST program to facilitate these regulatory activities.

2.7   Developing IST Programs for New Nuclear Power Plants

The nuclear industry has submitted applications for licenses to construct and operate new nuclear power plants. The NRC discusses policy and technical issues associated with new reactors, including the development of IST programs, in several Commission papers, such as the following:

- SECY-90-016, "Evolutionary Light Water Reactor (LWR) Certification Issues and their Relationship to Current Regulatory Requirements" (ADAMS Accession No. ML003707849)

- SECY- 93-087, "Policy, Technical, and Licensing Issues Pertaining to Evolutionary and Advanced Light-Water Reactor (ALWR) Designs" (ADAMS Accession No. ML003708021)

- SECY- 94-084, "Policy and Technical Issues Associated with the Regulatory Treatment of Non-Safety Systems in Passive Plant Designs" (ADAMS Accession No. ML003708068)

- SECY- 95-132, "Policy and Technical Issues Associated with the Regulatory Treatment of Non-Safety Systems (RTNSS) in Passive Plant Designs" (Accession No. ML003708005)

- Applicable Staff Requirements Memoranda (SRMs)

In a public memorandum dated July 24, 1995; the NRC staff consolidated the discussion of the policy and technical issues associated with the regulatory treatment of non-safety systems (RTNSS) in new passive plant designs provided in SECY-94-084 and SECY-95-132, and their associated SRMs.

The NRC regulations in 10 CFR Part 52, "Licenses, Certifications, and Approvals for Nuclear Power Plants," include design certifications for specific new reactor designs, such as the Advanced Boiling Water Reactor (ABWR) and AP1000 PWR. In addition, suppliers of design certifications have submitted applications for, or updates to, certification of designs for several new reactors. For example, the NRC is reviewing design certification applications for the Economic Simplified Boiling Water Reactor (ESBWR), Evolutionary Power Reactor (U. S. EPR), and U.S. Advanced Pressurized Water Reactor (US-APWR). The NRC has approved an amendment to the AP1000 certified design. The NRC regulations require new reactor suppliers to address the design of plant systems related to the performance of the IST program in their design certification application. While the design phase contains significant flexibility, new reactor vendors should design their plants to minimize the need for requests for relief from the IST provisions in the ASME OM Code. Under 10 CFR Part 52, the applicant has responsibility for the development of a plant-specific IST program for a combined operating license (COL) to construct and operate a nuclear power plant.

The Commission's SRM, dated September 11, 2002, for Commission Paper SECY-02-0067, "Inspections, Tests, Analyses, and Acceptance Criteria (ITAAC) for Operational Programs (Programmatic ITAAC)," (ADAMS Accession No. ML020700641) stated that ITAAC for an operational program are unnecessary if the COL application fully describes the program and its implementation and the NRC finds them to be acceptable. The Commission also stated that the burden is on the COL applicant to provide the necessary and sufficient programmatic information for approval of the COL without ITAAC.

In its May 14, 2004, SRM for SECY-04-0032, "Programmatic Information Needed for Approval of a Combined License Without Inspections, Tests, Analyses and Acceptance Criteria," (ADAMS Accession No. ML040230079) the Commission defined "fully described" as meaning that the program is clearly and sufficiently described in terms of the scope and level of detail to allow a reasonable assurance finding of acceptability. The Commission also noted that required programs should always be described at a functional level and at an increasing level of detail where implementation choices could materially and negatively affect the program effectiveness and acceptability. SECY-05-0197, "Review of Operational Programs in a Combined License

Application and Generic Emergency Planning Inspections, Tests, Analyses, and Acceptance Criteria," (ADAMS Accession No. ML060540043) summarizes the NRC position regarding the full description of operational programs to be provided by COL applicants. RG 1.206, "Combined License Applications for Nuclear Power Plants (LWR Edition)," provides guidance for COL applicants with respect to fully describing plant operational programs. The guidance in this NUREG may be used in developing and implementing the IST program for new nuclear power plants.

For a COL issued per 10 CFR Part 52, the NRC regulations in 10 CFR 50.55a(f)(4)(i) and 10 CFR 50.55a(g)(4)(i) state that inservice tests and inspections conducted during the initial 120-month interval to verify operational readiness of applicable plant components, whose function is required for safety, must comply with the requirements in the latest edition and addenda of the ASME Code, incorporated by reference in 10 CFR 50.55a(b) on the date 12 months before the date scheduled for initial fuel loading (or the optional ASME Code cases listed in RG 1.192) subject to the conditions listed in 10 CFR 50.55a. As discussed in RIS 2012-08, "Developing Inservice Testing and Inservice Inspection Programs under 10 CFR Part 52," dated July 16, 2012, (ADAMS Accession No. ML112114A010) licensees may submit a request under 10 CFR 50.55a to apply the edition and addenda of the ASME OM Code specified in the COL application for the initial 120-month IST interval as an alternative to the latest edition and addenda of the ASME OM Code, where the differences between the IST provisions in these Code editions and addenda are addressed.

Several applications under 10 CFR Part 52 have been submitted for COLs to construct and operate new nuclear power plants that reference certified reactor designs or designs under certification review. In addition to addressing design aspects related to the IST program, new reactor design certification applicants typically provide a description of generic aspects of the IST program to allow the COL applicants to incorporate by reference this design certification information in their COL application. The NRC staff reviews the description of the IST program in the COL application, with its incorporation by reference of IST provisions in the applicable design certification documentation, as part of the safety evaluation for the COL application. The NRC staff will conduct inspections of the development and implementation of the IST program following COL issuance.

ASME has a program underway to establish improved IST provisions in the ASME OM Code for pumps, valves, and dynamic restraints to be used in new reactors. ASME has prepared a White Paper that discusses its plans to update the ASME OM Code for new reactors. For example, the ASME White Paper identifies lessons learned from operating experience at current nuclear power plants, and from research sponsored by the nuclear industry and the NRC, that are applicable to IST programs for new reactors. The ASME White Paper also identifies new reactor issues that can affect IST programs to be developed for new reactors.

Lessons learned from nuclear power plant operating experience and research that should be considered in the development of IST programs for new reactors include, for example:

1. Design and qualification of pumps, valves, and snubbers to allow IST activities (including sufficient flow testing) to assess the operational readiness of those components, including ASME Standard QME-1-2007, "Qualification of Active

Mechanical Equipment Used in Nuclear Power Plants," as accepted in Revision 3 to RG 1.100, "Seismic Qualification of Electric and Active Mechanical Equipment and Functional Qualification of Active Mechanical Equipment for Nuclear Power Plants," to incorporate lessons learned in the qualification of mechanical equipment for nuclear power plants.

2. Performance and testing of MOVs that indicate the need for improved MOV activities, such as importance of adequate design and qualification, sufficient flow during testing to assess valve performance, consideration of MOV performance parameters (including valve disk and stem friction coefficients, reduced voltage, elevated temperature, and load sensitive behavior), use of adequate diagnostic instrumentation to allow proper evaluation and setup, improved maintenance and personnel training, monitoring of potential motor magnesium rotor degradation, and justification for motor control center testing.

3. Application of MOV lessons learned to other Power-Operated Valves (POVs).

4. Provisions for bi-directional testing of check valves.

5. Implementation of pre-service testing (PST) and comprehensive pump testing (CPT) provisions without the need for relief from the ASME OM Code provisions.

6. Consideration of potential adverse flow effects on plant components from flow-induced vibration resulting from hydrodynamic loads and acoustic resonance.

New reactor issues that should be considered in the development of IST programs for new nuclear power plants include, for example:

1. Description of IST programs by COL applicants in accordance with 10 CFR Part 52 with implementation of design certification provisions for design, qualification, and IST activities.

2. Coordination of PST and ITAAC so that testing is performed once for both purposes. For example, implementation of PST requirements and the new 10 CFR Part 52 ITAAC closure and maintenance process both need to be accomplished. Under the new 10 CFR Part 52 process, an applicant is required to meet OM Code requirements after a 10 CFR 52.103(g) finding is made, although it would be preferable to complete the PST requirements earlier.

3. Design, qualification, and IST and inspection activities for pyrotechnic-actuated (squib) valves that have high safety significance, and that represent more significant engineering challenges for new reactors than for current operating plants.

4. Design of plant systems and development of IST programs to minimize the need for relief from the ASME OM Code provisions.

5. Design, qualification, PST, and IST activities for regulatory treatment for non-safety systems (RTNSS) components that perform safety significant functions.

6. Development and implementation of risk-informed IST programs, including programs under 10 CFR 50.69, "Risk-informed Categorization and Treatment of Structures, systems, and Components for Nuclear Power Plants," for new reactors.

7. Consideration of appropriate Code and standard modifications for design, qualification, PST and IST activities in response to application of software-based digital technology in mechanical components (e.g., pumps and valves).

Applicants for new nuclear power plants should consider the information in this NUREG and other sources, such as the ASME program to update the OM Code, in developing their IST programs.

As part of its initial evaluation of the PST and IST provisions for new reactors, ASME has prepared PST and IST provisions for squib valves in new reactors that are included in the 2012 Edition of the ASME OM Code. The NRC staff will consider these PST and IST provisions for squib valves in the next rulemaking to update 10 CFR 50.55a to incorporate by reference the 2012 Edition of the ASME OM Code.

**Table 2.1 Typical Systems and Components in an Inservice Testing Program
for a Pressurized-Water Reactor**

Typical safety-related, Code-class system in pressurized-water reactors	Typical components in an inservice testing program
Reactor coolant system and flowpaths for establishing natural circulation	Power-operated relief valves and associated block valves Reactor high point and head vents Primary system safety and relief valves (pressurizer Code safety valves) Valves in any proposed flowpath used for long-term core cooling or safe shutdown Pressure boundary isolation valves Valves in lines to pressurizer relief/quench tank
Main steam system	Main steam isolation valves (MSIVs) Main steam non-return valves (if applicable) Secondary system safety and relief valves Atmospheric dump valves Auxiliary feedwater turbine steam supply valves Steam generator blowdown isolation valves
High-pressure safety injection system	High-pressure injection pumps and discharge check valves Injection valves in injection flowpath Isolation valves Valves for the refueling water storage tank (RWST) borated water storage tank (BWST), and refueling water tank (RWT), including vacuum breakers
Chemical and volume control or makeup system	Charging or makeup pumps and suction/discharge check valves Valves in charging/makeup flowpath Boric acid transfer pumps and suction/discharge check valves Valves in emergency boration flowpaths Relief valves

Table 2.1 Typical Systems and Components in an Inservice Testing Program for a Pressurized-Water Reactor (continued)

Typical safety-related, Code-class system in pressurized-water reactors	Typical components in an inservice testing program
Low-pressure safety injection system	Injection pumps and suction/discharge check valves Valves associated with safety injection accumulators and core flood tanks Recirculation flowpath valves, including containment sump isolation valves Isolation valves (high-low pressure interface) Relief valves
Shutdown cooling, residual heat removal, or decay heat removal systems	Pumps and suction/discharge check valves Valves in flowpath Isolation valves (high-low pressure interface) Relief valves
Containment spray system	Containment spray pumps and suction/discharge check valves Valves in flowpaths to spray header Isolation valves Valves in spray additive flowpath Spray additive tank valves, including vacuum breakers
Main feedwater system	Main feedwater isolation valves
Auxiliary feedwater system	Auxiliary feedwater pumps and suction/discharge check valves Valves in flowpath to steam generators Valves in suction lines Valves between normal and ultimate heat sink suction sources Relief valves and isolation valves

**Table 2.1   Typical Systems and Components in an Inservice Testing Program
for a Pressurized-Water Reactor (continued)**

Typical safety-related, Code-class system in pressurized-water reactors	Typical components in an inservice testing program
Primary containment system	Containment isolation valves (various systems) Containment combustible gas venting valves Containment atmosphere sampling valves (if within the scope of 10 CFR 50.55a)
Component cooling water system	Component cooling water pumps and discharge check valves Valves in letdown cooling water flowpath Valves in reactor coolant pump seal injection and cooling water flowpath Valves needed to isolate a rupture of the thermal barrier Relief valves
Spent fuel pool/pit cooling system	Spent fuel cooling pumps and suction/discharge check valves Valves in flowpath from ultimate heat sink source supply
Service water system	Service water pumps and suction/discharge check valves Valves in flowpath to auxiliary feedwater system Valves in flowpaths to emergency room coolers Valves in flowpaths to containment emergency coolers Valves in flowpaths to emergency diesel generator heat exchangers Isolation and cross-tie valves Valves in ultimate heat sink source flowpaths Valves in standby or backup service water, if applicable

Table 2.1   Typical Systems and Components in an Inservice Testing Program for a Pressurized-Water Reactor (continued)

Typical safety-related, Code-class system in pressurized-water reactors	Typical components in an inservice testing program
Emergency diesel generator system (within scope of 10 CFR 50.55a)	Fuel oil storage and transfer pumps and valves Diesel generator external cooling (service water) Engine air start check valves Air receiver relief valves
Ventilation systems	Pumps and valves in control room emergency cooling water supply flowpath
Instrument air system (if within the scope of 10 CFR 50.55a)	Air supply to containment purge valves Air supply to power-operated relief valves (PORVs) Air supply to MSIVs

**Table 2.2   Typical Systems and Components in an Inservice Testing Program
for a Boiling-Water Reactor**

Typical safety-related, Code-class system in boiling-water reactors	Typical components in an inservice testing program
Nuclear boiler and reactor recirculation system	Primary system isolation valves Excess flow check valves
Main steam system	MSIVs and actuator valves (pilot valves, accumulator check valves) Main steam safety and relief valves Main steam safety valve discharge rupture diaphragm valve MSIV leakage valves
High-pressure core coolant injection (HPCI) system	Pump and suction/discharge check valve Valves in injection flowpath Isolation valves, including valves in test lines Excess flow check valves HPCI pump turbine valves, including turbine exhaust vacuum breakers (unless considered skid-mounted)
High-pressure core spray system	Pumps and suction/discharge check valves Valves in injection flowpath Isolation valves, including valves in test lines
Reactor core isolation cooling (RCIC) system (if safety-related)	Pump and suction/discharge check valve RCIC pump turbine valves Excess flow check valves Isolation valves
Reactor water cleanup system	Containment isolation valves

**Table 2.2   Typical Systems and Components in an Inservice Testing Program
for a Boiling-Water Reactor (continued)**

Typical safety-related, Code-class system in boiling-water reactors	Typical components in an inservice testing program
Residual heat removal (RHR) system	RHR pumps and suction/discharge check valves Isolation and cross-tie valves Pump suction relief valves RHR heat exchanger thermal relief valves Valves in injection flowpath Flow control valves
Spent fuel pool cooling system	Fuel pool pumps and suction/discharge check valves Ultimate heat sink supply valve
Feedwater coolant injection and isolation condenser system (if applicable)	Reactor feedwater pumps and suction/discharge check valves Condensate pumps and suction/discharge check valves Condensate booster pumps and suction/discharge check valves Emergency condensate transfer pump and suction/discharge check and isolation valves Isolation and bypass valves Vent valves Makeup to condenser shell check valves
Standby liquid control (SBLC) system	SBLC pumps and suction/discharge check valves Relief valves Injection line valves Explosively-actuated squib valves
Main feedwater system	Isolation valves

**Table 2.2 Typical Systems and Components in an Inservice Testing Program
for a Boiling-Water Reactor (continued)**

Typical safety-related, Code-class system in boiling-water reactors	Typical components in an inservice testing program
Primary containment system	Containment isolation valves including excess flow check valves (various systems) Containment atmosphere monitoring system valves Containment atmosphere dilution system valves Containment pressure suppression and vents
Closed cooling or component cooling water system	Pumps and suction/discharge check valves Valves in flowpaths to safety-related coolers
Service water system	Pumps and suction/discharge check valves Isolation and cross-tie valves Valves in flowpaths to safety-related coolers Valves in flowpaths to diesel generator coolers Valves in standby or backup service water Valves in flowpath from ultimate heat sink source Valves in residual heat removal service water flowpath
Control rod drive system (portions within the scope of 10 CFR 50.55a)	Scram dump valves Scram discharge volume vent valves Scram discharge volume drain valves Accumulator rupture disks Hydraulic control unit control valves Drive water backflow prevention valves
Emergency diesel generator systems (if within the scope of 10 CFR 50.55a)	Fuel oil storage and transfer pumps and valves Diesel generator external cooling (service water) Engine air start check valves Air receiver relief valves

**Table 2.2   Typical Systems and Components in an Inservice Testing Program
for a Boiling-Water Reactor (continued)**

Typical safety-related, Code-class system in boiling-water reactors	Typical components in an inservice testing program
Ventilation systems	Pumps and valves in control room emergency cooling water supply flowpath
Instrument air system (if within the scope of 10 CFR 50.55a)	MSIV accumulator check valves MSIV pilot valves Automatic depressurization system (ADS) valve accumulator check valves ADS pilot valves
Traversing incore probe system (if within the scope of 10 CFR 50.55a)	Containment isolation valves

Table 2.3 Example Data Table for Pumps

PLANT NAME/UNIT

PUMP TESTING PLAN

Revision 3
Date: 1-15-03
Page: 1 of 3

Pump List					Parameters				
SYSTEM	PUMP ID	P&ID NO.	COORD.	PUMP GROUP	S	P	dp	Flow	V (PR-1)
Residual Heat Removal	RHR-01	M-402, Sh. 1	D-4	A	(1)	(2)	Q/2Y	Q/2Y	Q/2Y
	RHR-02	M-402, Sh. 2	G-4		(1)	(2)	Q/2Y	Q/2Y	Q/2Y
	RHR-03	M-402, Sh. 2	F-5		(1)	(2)	Q/2Y	Q/2Y	Q/2Y
Auxiliary Feedwater	AFW-01	M-408, Sh 1	B-5	B	(1)	(2)	Q/2Y	Q/2Y	2Y
	AFW-02	M-408, Sh. 1	B-8		(1)	(2)	Q/2Y	Q/2Y	2Y
	AFW-03	M-408, Sh. 1	B-11		Q/2Y	(2)	Q/2Y	Q/2Y	2Y
Service Water	SWS-01	M-335, Sh 1	F-9	A	(1)	(2)	Q/2Y	PR-3	PR-2
	SWS-02	M-335, Sh 2	D-4		(1)	(2)	Q/2Y	PR-3	PR-2
	SWS-03	M-335, Sh. 3	E-8		(1)	(2)	Q/2Y	PR-3	PR-2
	SWs-04	M-335, Sh. 4	C-4		(1)	(20	Q/2Y	PR-3	PR-2
Standby Liquid Control	SLC-01	M-367, Sh. 1	D-9	B	(1)	2Y	(3)	Q/2Y	2Y
	SLC-02	M-367, Sh. 1	D-4		(1)	2Y	(3)	Q/2Y	2Y

Note (1): Pump is directly coupled to a **content** speed synchronous or induction type driver
Note (2): Discharge pressure is a required parameter for positive displacement pumps only.
Note (3)" dP is not a required parameter for positive displacement pumps.

Legend:

	S	Speed
	P_i	Pressure
	dP	Differential Pressure
	PR	Pump Relief Request
	Q	Quarterly
	V	Vibration

Table 2.4 Useful Abbreviations for Valve Data

Parameter	Abbreviation	Description
Valve Type	GT	Gate valve
	GB	Globe valve
	CK	Check valve
	RV	Relief valve
	SC	Stop check
	BF	Butterfly valve
	DI	Diaphragm valve
	EX	Explosive valve
	BA	Ball valve
Actuator Type	MO	Motor-operated
	SO	Solenoid-operated
	AO	Air-operated
	HO	Hydraulic-operated
	SA	Self-actuated
	MA	Manual
	PA	Pilot-actuated
Safety Position(s)	O	Open
	C	Closed
	O/C	Both open and closed
	T	Throttled
Tests(s) Performed	FS	Full-stroke exercise valve to safety position(s)
	PS	Part-stroke exercise valve
	LT	Leak-rate test valve to Section XI requirements
	LJ	Leak-rate test valve to Appendix J requirements
	ST	Measure the full-stroke times of the valve
	FT	Observe the fail-safe operation of the valve
	PI	Verify the valve position indication
	RV	Safety and relief valve test
	EX	Explosive valve test
Test Frequency	Q	Test performed once every 92 days
	CS	Test performed during cold shutdowns, but not more frequently than once every 92 days
	RF	Test performed each reactor refueling outage
	2Y	Test performed once every 2 years
	RV	Test relief valve at OM schedule
	SD	Disassemble, inspect, and manually exercise one valve from specified group each reactor refueling outage

Figure 2.1
FLOW CHART – Development of Inservice Testing Program for Pumps & Valves*

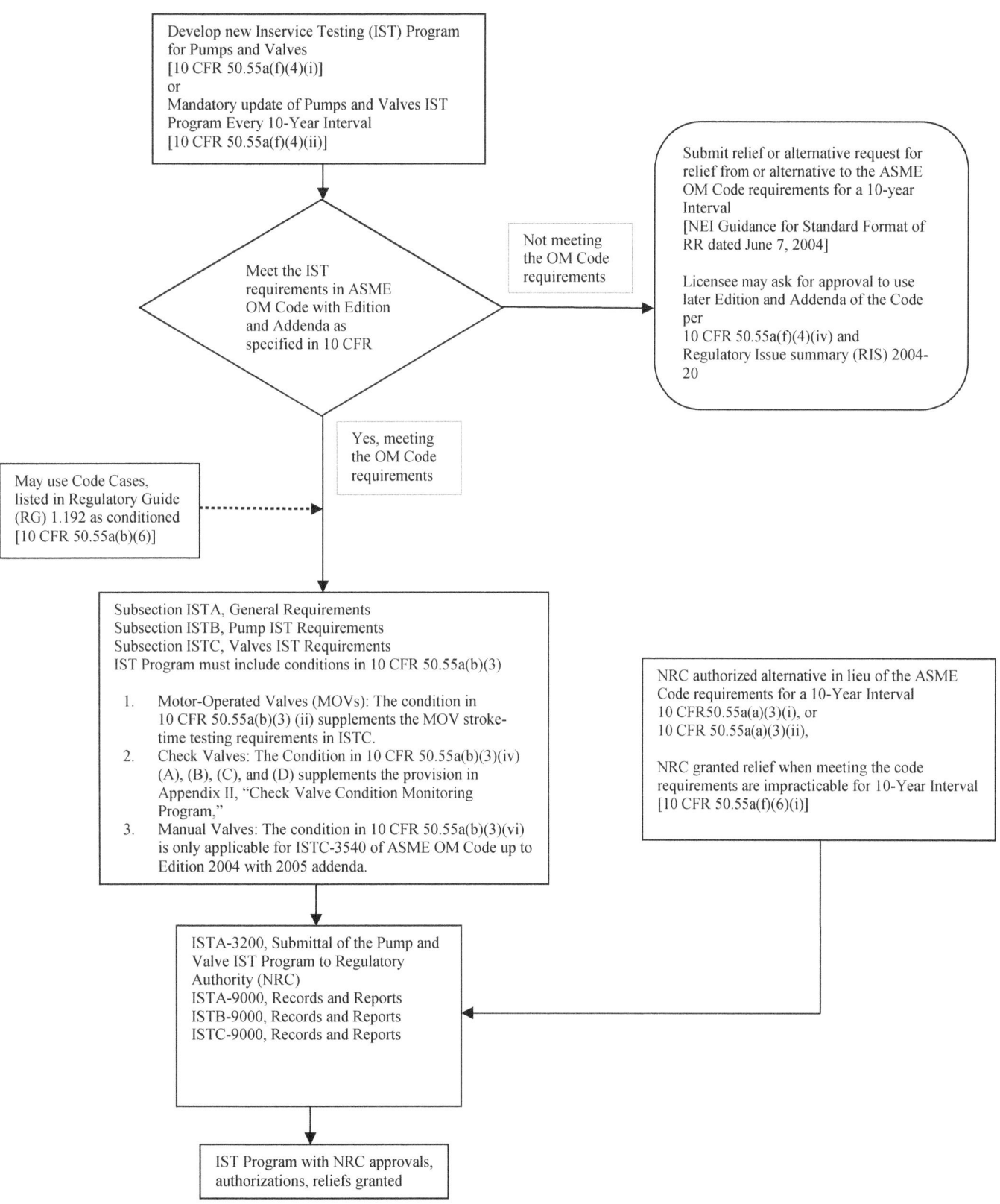

*Note: Flow chart provided for guidance only. For complete details see 10 CFR 50.55a

3. GENERAL GUIDANCE ON INSERVICE TESTING

3.1 Inservice Test Frequencies and Extensions for Valve Testing

The ASME OM Code generally specifies quarterly testing of valves. Subsection ISTC of the Code allows licensees to defer valve exercising to cold shutdown or refueling outages if it is not practical to exercise the valves during plant operation. Impractical conditions justifying test deferrals are those that could result in an unnecessary plant shutdown, cause unnecessary challenges to safety systems, place undue stress on components, cause unnecessary cycling of equipment, or unnecessarily reduce the life expectancy of the plant systems and components. Table 3.1 (below) lists the tests and associated test frequencies required by the Code.

Table 3.1 ASME OM Code Tests and Test Frequencies for Pumps and Valves

Test	Frequency
Measure pump parameters	Once every 3 months (Group A, B) Biennially (Comprehensive Test) Exceptions: · Pumps in systems that are out-of-service · Group B Pumps lacking required fluid inventory
Exercise Category A and B valves	Once every 3 months Exceptions: · Extension because of impracticality · Valves in regular use · Valves in systems out of service
Measure stroke times of power-operated Category A and B valves	Once every 3 months Exceptions: · Extension because of impracticality · Valves in regular use · Valves in systems out of service
Verify remote position indication	Once every 2 years
Observe operation of fail-safe actuators for applicable valves	Once every 3 months, except for extension because of impracticality
Leak-test Category A and A/C valves	Once every 2 years
Test safety and relief valves, primary containment vacuum relief valves, and non-reclosing pressure relief devices	Test interval specified by OM Appendix I
Exercise check valves	Once every 3 months Exceptions: · Extension because of impracticality · Valves in regular use · Valves in systems out of service
Test explosively actuated valves	20 percent tested once every 2 years. Charges shall not be older than 10 years.

3.1.1 Deferring Valve Testing to Each Cold Shutdown or Refueling Outage

The OM Code allows licensees to test valves during cold shutdowns if it is impractical to test the valves quarterly during plant operation. Subsection ISTC-3500 provides guidance and alternatives. Therefore, exercising valves during cold shutdown outages does not constitute a deviation from the OM Code and does not require a relief request if the licensee determines that quarterly testing is impractical. Similarly, the OM Code allows licensees to test valves during each refueling outage if it is impractical to test the valves during cold shutdowns. In such instances, the licensee should identify the valves for which testing is deferred and the inservice testing (IST) program document should specify the basis for determining that quarterly and/or cold shutdown testing is impractical.

In the past, the NRC staff has provided examples of valves that be excluded from exercising (cycling) tests during plant operations. The excluded valves include the following examples:

(1)   All valves that would cause a loss of system function if they were to fail in a nonconservative position during the cycling test. Valves in this category would typically include all non-redundant valves in lines such as a single discharge line from the refueling water storage tank (RWST) or accumulator discharge lines in PWRs and the HPCI turbine steam supply and HPCI pump discharge in BWRs. Other valves may fall into this category under certain system configurations or plant operating modes. For example, when one train of a redundant system [such as an emergency core cooling system (ECCS)] is inoperable, non-redundant valves in the remaining train should not be cycled because their failure would cause a loss of total system function.

(2)   All valves that would result in a loss of containment integrity if they failed to close during a cycling test. Valves in this category would typically include all valves in containment penetrations where the redundant valve is open and inoperable.

(3)   All valves that, when cycled, could subject a system to pressures in excess of their design pressures. For the purpose of a cycling test, it is assumed that one or more of the upstream check valves has failed unless positive methods are available for determining the pressure or lack thereof on the high-pressure side of the valve to be cycled. Valves in this category would typically include the isolation valves of the residual heat removal/or shutdown cooling system and, in some cases, certain ECCS valves

The guidance in this NUREG and in the NRC's letters issued in 1976 to licensees do not supersede the TS requirements.

A licensee may request relief from quarterly testing where such testing would impose a hardship (e.g., entering a limiting condition for operation of 3 to 4 hours in duration, repositioning a breaker from "off" to "on," and necessitating for manual operator actions to restore the system if an accident were to occur while the test was in progress). For such situations, the risk associated with quarterly testing may outweigh the benefits that might otherwise be achieved. (Section 3.1.2 gives guidance on these types of situations.) Thus, it is appropriate for licensees to weigh the safety impact against the benefits of testing as a basis for deferring testing from a quarterly frequency to cold shutdown or refueling outages. NUREG/CR-5775, "Quantitative Evaluation of Surveillance Test Intervals Including Test-Caused Risks," dated February 1992, (ADAMS Accession No. ML027410457) describes a method for making this comparison.

The following sections discuss issues related to deferring valve testing. These sections do not apply to testing that is required following maintenance or repair activities.

3.1.1.1 IST Cold Shutdown Testing

Although Subsection ISTC of the ASME OM Code does not include schedules for cold shutdown testing, an acceptable method is to ensure that the valves tested in the preceding cold shutdown are the last valves tested during the next cold shutdown, with the exception of valves that must be tested during each cold shutdown. The following is a sample schedule for 15 cold shutdown tests:

-
- First cold shutdown: Complete Tests 1, 2, 3, 4, 5, and 6
- Second cold shutdown: Complete Tests 7, 8, 9, and 10
- Third cold shutdown: Complete Tests 11, 12, 13, 14, 15, 1, 2, and 3
- Fourth cold shutdown: Complete Tests 4, 5, 6, and 7

Subsection ISTC-3520 discusses exercising valves during both plant operation and cold shutdown as circumstances and situations apply. While the discussion does not specifically address testing in hot standby or hot shutdown, valves should be exercised in the appropriate mode of operation. For a valve that cannot be tested in operation, testing might be practical during hot standby, hot shutdown, cold shutdown, or a refueling outage.

Valves that must be operable during cold shutdown may be tested during plant operation in accordance with Subsection ISTC-3520, "Exercising Requirements," or Subsection ISTC-3550, "Valves in Regular Use." By contrast, Subsection ISTC-3550 applies if the component's "normal use" is during cold shutdown.

3.1.1.2 Testing at a Refueling Outage Frequency for Valves Tested During Power Ascension

Subsection ISTC-3520 specifies that valves that are tested on a refueling outage frequency should be tested before returning the plant to operation at power. Several licensees have indicated that they cannot test certain valves until power ascension begins. The NRC staff has included this section to provide guidance for such valves and to indicate that the operability TS would control the timing for testing such valves. It is intended that the IST program document will identify such valves as being tested on a refueling outage frequency, even though the plant may actually return to "operation" at power before the testing is completed. A similar intent applies to valves that are tested during power ascension from cold shutdowns (which are not refueling outages); however, Subsection ISTC uses different language in discussing valves that are tested on a cold shutdown frequency.

Before beginning power ascension from a refueling outage, licensees normally complete the tests of those valves that are tested during each refueling outage. However, for valves that can only be tested during power ascension or at power, licensees may begin increasing the power level and changing modes in accordance with TS requirements and may test the applicable valves when plant conditions allow testing. This situation also could apply to valves that are tested during power ascension or at power following a cold shutdown outage.

NRC Recommendation

The ASME OM Code Subsection ISTC-3520 requires that all valves testing scheduled for performance during a refueling outage shall be completed before returning the plant to operation. For valves which can only be tested during power ascension, TS requirements (for the valves or the associated system) determine when the valves are required to be operable. The testing for these valves may be scheduled for refueling outages or during cold shutdown conditions, but completed during power ascension from the refueling outages.

The NRC staff has determined that testing of such valve during plant startup period but prior to reaching steady state full power operation is acceptable without a relief request, provided that the licensee meets all requirements of ISTC-3520.

Basis for Recommendation

The staff has determined that the guidance in this section is consistent with ASME OM Code Subsection ISTC-3520 and the TS requirements and, therefore, is acceptable for meeting these provisions.

3.1.1.3 De-Inerting Containment of Boiling-Water Reactors to Allow Cold Shutdown Testing

According to 10 CFR 50.44, "Standards for Combustible Gas Control System in Light-Water-Cooled Power Reactors," each BWR that is equipped with a Mark I or Mark II containment must have provisions for an inerted containment atmosphere during power operation to protect against a burn or explosion of hydrogen gas generated by the core metal-water reaction following a postulated loss-of-coolant accident (LOCA).

Licensees regularly monitor oxygen content in the containment atmosphere during normal power operation, and the plant's TS specify the maximum oxygen concentrations allowed. Since hydrogen generation is not a concern during cold shutdown or refueling outages, the TS allow the containment atmosphere to be de-inerted. However, licensees do not routinely de-inert the containment during cold shutdown outages because of impracticality concerns associated with the time needed to de-inert and re-inert the containment, and because of the amount of nitrogen necessary for inerting.

For certain valves that are located in the inerted containment, Subsection ISTC-3500 allows licensees to perform testing during cold shutdown outages because it is not practical to test such valves during power operation. The staff has determined that it is impractical to de-inert the containment during each cold shutdown outage solely to perform such routine testing or repair activities.

NRC Recommendation

The staff considers it impractical to de-inert the containment merely to conduct regularly scheduled valve testing, and the OM ASME Code allows licensees to defer such testing to a refueling outage when the containment is de-inerted for refueling or other reasons. The staff also has determined that few outages require de-inerting, and it is unnecessary to maintain a separate schedule for valve testing. Consequently, testing is at the discretion of the licensee in the event of an extended cold shutdown that necessitates de-inerting the containment. Factors to be considered in the licensee's decision-making might include the length of the shutdown and the extent of other outage activities. In addition, for extended outages that last several months, the requirements of Subsection ISTC-3570 may apply for valves in systems that are out-of-service.

Basis for Recommendation

Subsection ISTC allows licensees to extend the test interval to defer valve testing to refueling outages if such testing is impractical at quarterly intervals (during power operations) or during cold shutdown outages. Consequently, it is also acceptable for licensees to extend the test interval for those valves which cannot be tested unless the containment is de-inerted.

Unless the licensee has some other reason to enter the drywell during cold shutdown outages, the staff regards it as impractical to de-inert the drywell during such outages merely to perform valve testing. The staff's position is based on the time and effort needed to de-inert, re-inert, and replace lost nitrogen gas (which could delay the plant's return to power operation).

3.1.1.4 Stopping Reactor Coolant Pumps for Cold Shutdown Valve Testing

Licensees frequently defer the testing of certain valves in support systems that perform functions that are vital to the continued operability of the reactor coolant pumps, such as component cooling and the supply and return of seal water during cold shutdown. Exercising these valves while the reactor coolant pumps are operating could result in pump damage, and stopping the pumps could extend the cold shutdown period.

NRC Recommendation

The staff recommends that licensees test the affected valves on a refueling outage schedule and during plant outages when the reactor coolant pumps are stopped for a sufficient period of time, but not more often than once every 92 days.

Basis for Recommendation

Subsection ISTC of the ASME OM Code allows licensees to extend the test interval to defer testing to refueling outages when it is not practical to perform the tests during power operation or cold shutdown outages. The NRC staff believes that licensees need not schedule valve testing that requires stopping and restarting reactor coolant pumps during each cold shutdown solely to allow for the testing of such valves. This repetitive cycling would increase pump wear and stress, as well as the number of cycles of related plant equipment, and could extend the length of cold shutdown outages. Consequently, licensees may consider establishing a schedule to account for extended cold shutdown outages that would allow for valve testing when the reactor coolant pumps are stopped for a sufficient period of time. However, valves are to be tested at least during each refueling outage, as required by Subsection ISTC.

3.1.2   Entry into a Limiting Condition for Operation (LCO) To Perform Testing

ASME OM Code Subsections ISTB and ISTC allow deferred testing until a cold shutdown or refueling outage, if testing is not practicable at power. The staff believes that it is better to defer testing as allowed by the Code, rather than entering into an LCO ACTION to perform IST or requesting approval of a relief or exemption to perform such IST without entering into LCO ACTION. See Section 3.1.3 for guidance on scheduling of inservice testing.

3.1.3   Scheduling of Inservice Tests

The ASME OM Code requires that testing is to be performed normally within a certain time periods.

NRC Recommendation

To eliminate ambiguity concerning the periods stated in the OM Code, the staff recommends that licensees use the stated test frequency definitions (as shown in Table 3.2, below). For example, Subsection ISTC-3510 requires licensees to test Category A and B valves "nominally every 3 months." For quarterly testing, the staff recommends that licensees schedule the pump and valve tests such that a particular test is performed at approximately the same time within each quarter. For example, if a test procedure applies to many valves and, thus, requires 2 to 3 weeks or more to complete, the licensee would typically begin the procedure at approximately the same time in each quarter and include directions to perform tests in a specified order to ensure that specific valves are tested "at least once every 92 days."

Table 3.2 ASME OM Code Terms for Inservice Testing Activities

Term	Required frequency for IST activities (at least once every)
Monthly	31 days
Quarterly (or Every 3 months)	92 days
Yearly (or Annually)	366 days
Refueling	refueling outage
2 years	24 months

In the past, the technical specifications defined that licensees perform each applicable test within the specified time interval, with a maximum allowable extension not to exceed 25 percent of the test interval. Licensees could not extend the test intervals for safety and relief valves defined in Appendix I, "Inservice Testing of Pressure Relief Devices in Light-Water Reactor Nuclear Power Plants," to the OM Code, other than to coincide with a refueling outage.

The Code specifies performing the tests throughout extended shutdown periods for equipment that must be returned to operable status. Most equipment must be tested before being returned to service after being out-of-service for an extended period in accordance with TS requirements (if applicable). The OM Code provisions in Subsections ISTB-3420 and ISTC-3570 specify that licensees need not follow the test schedule if the system in which the component was installed was declared inoperable or was not required to be operable. However, this applies only if the component was not out-of-service for repair or replacement. For repair or replacement, the component must be tested within 3 months of the system being returned to service.

Basis for Recommendation

This recommendation is based on the standard technical specifications, which the NRC staff have developed, reviewed, and approved as staff technical positions. The specified intervals and extensions apply directly to IST of pumps and valves, as applicable. In Interpretation XI-78-01, the ASME Code Committee clarified the intent of the "2-year" frequency specified for verifying position indication and performing leak rate testing, stating that the Code test and examination frequency relates to periods of time, rather than refueling outages. The Code references refueling outages to preclude the necessity to shut down the plant solely for IST. The OM Code specifies that licensees must perform the valve position indicator test and leak rate test at least every 2 years, without regard to the frequency of refueling outages.

The NRC staff recommendation for testing during extended shutdown periods is consistent with TS and Code requirements, whichever are more conservative. Responding to inquiry

IN 92-025A, the ASME Code Committee stated that Subsection ISTC-3510 intends that testing be conducted every 3 months, including during extended shutdown periods, for valves other than those declared inoperable in accordance with ISTC-3570.

The OM Committee made a similar clarification in OM Interpretation 93-1, stating that it is intended that testing be conducted every 3 months, including during extended periods, for valves other than those declared inoperable or not required to be operable. The ASME Code committee prepared Code Case OMN-20 to address this issue of test interval and 25 percent margin.

3.2   Reserved

3.3   120-Month Updates Required by 10 CFR 50.55a(f)(4)(ii)

10 CFR 50.55a(f)(4)(ii) requires licensees to revise their IST programs every 120 months to reflect the latest edition and addendum to the OM Code incorporated by reference into 10 CFR 50.55a(b)(3) that is in effect 12 months before the start of the new 120-month IST interval.

After the initial 120-month interval, in accordance with 10 CFR 50.55a(f)(4)(ii), licensees must conduct inservice tests during successive 120-month intervals to verify the operational readiness of pumps and valves within the scope of the ASME Code. In conducting these inservice tests, licensees must comply with the provisions of the latest edition and addenda of the Code incorporated by reference in 10 CFR 50.55a(b) that is in effect 12 months before the start of the 120-month IST interval, subject to the conditions listed in paragraph (b). In addition, 10 CFR 50.55a(f)(5)(iv) specifies that where a pump or valve test requirement by the Code or addenda is determined to be impractical by the licensees and is not included in the new IST program interval, such that the basis for this determination must be submitted for NRC review and approval not later than 12 months after the expiration of the initial 120-month interval of operation from the start of the facility commercial operation and each subsequent 120-month interval of operation during which the test is determined to be impractical.

3.3.1   Extension of Interval

The IST interval may be extended in accordance with Subsection ISTA-3120(d):

> Each IST interval may be extended or decreased by as much as 1 year. Adjustments shall not cause successive intervals to be altered by more than 1 year from the original pattern of intervals.

Subsection ISTA-3120(e) further states that, for units that are continuously out-of-service for 6 months or more, licensees may extend the IST interval during which the outage occurred for a period equivalent to the outage, and may extend the original pattern of intervals accordingly for successive intervals.

NRC Staff Recommendation

Licensees must establish the next updated program to the latest edition of the Code incorporated in the regulation 12 months before the new date. For example, if a licensee has an extension from December 14, 2010, to September 16, 2011, in accordance with the Code, the licensee's program for the new interval must meet the edition of the Code incorporated in 10 CFR 50.55a(b) as of September 16, 2010. When extending its 10-year IST interval by as much as 1 year, as allowed by ISTA-3120(d), licensees may continue to apply the same Code edition and addenda from its current 120-month interval during this extended 1-year period. The staff recommends that licensees should inform the NRC of any extension before the date that would have been the end of the current interval. An extension beyond 1 year (other than for extended outages, as permitted by ISTA-3120(e)) requires NRC approval of an alternative to or exemption from the Code provisions of 10 CFR 50.55a, "Codes and Standards," as applicable.

Basis for Recommendation

While it is not mandatory to maintain identical intervals for inservice inspection (ISI) and IST, it is often desirable in order to maintain the same edition of the Code for all plant activities related to ISI and IST. Even though 10 CFR 50.55a does not discuss extending the intervals, the Code is incorporated by reference in the regulation and, therefore, has the same effect as the regulation. Although NRC approval is not required for 1-year extensions of the interval, licensees would avoid any discrepancies in the interval dates by informing the NRC of the extension and documenting it in the IST program document. Because the Code does not allow extension beyond 1 year (other than for extended outages), such an extension would require NRC approval of an alternative to or exemption from the Code provisions to comply with the regulatory requirements.

3.3.2 Concurrent Intervals

Several licensees have established concurrent intervals for all units at sites with multiple units, so that each unit is updated to a newer edition of the Code at the same starting date. Because the regulations do not specifically allow for adjustments to accommodate concurrent intervals among multiple units, when the interval start dates are to be concurrent, licensees may request a one-time alternative to or exemption from 10 CFR 50.55a or the Code, as applicable. If a licensee prefers not to request an alternative or exemption, the establishment of concurrent intervals would require that the licensee must perform program updates for a particular unit more often than once every 120 months. 10 CFR 50.55a(f)(4)(iv) permits IST of pumps and valves to meet the provisions in subsequent Code editions and addenda (or portions thereof) that are incorporated by reference in 10 CFR 50.55a(b), subject to the conditions listed, and subject to NRC approval. This regulation allows licensees to update their programs before the end of a 120-month interval with NRC approval.

NRC Recommendation

If a licensee elects to use the same Code edition for multiple units, the licensee must request an alternative to the Code or exemption, to extend a unit's interval by more than 1 year in order to place multiple units on a concurrent interval for IST. To establish concurrent intervals without an alternative or exemption, the licensee must update the referenced edition of the Code more frequently for the selected unit(s) to remain in compliance with 10 CFR 50.55a, except in the case where the interval dates are within 12 months, whereby the Code allowance for an extension would result in concurrent intervals. The NRC will likely grant relief under

10 CFR 50.55a(f)(4)(iv) to allow the licensee to update to later editions of the Code, if the licensee uses the following guidelines:

(1)   Without obtaining an alternative or exemption, the licensee may perform the IST program for multiple units using the same edition of the Code at concurrent intervals if the initial interval for combining the programs is established such that no single unit is tested at an interval of more than 120 months (or no greater than the interval extension allowed by the Code). Thus, the licensee must use the interval for the first unit that was licensed for commercial operation to establish the interval dates and establish the correct Code edition according to the most recent required for either unit.

(2)   To exceed 120 months, other than as addressed in the Code for an extension, the licensee must first obtain approval of an alternative to the Code or exemption from 10 CFR 50.55a unless the licensee intends to repeatedly update both units more often than the required 120 months. Under such an approach, the licensee would test each unit according to the most recent edition of the Code required for either unit.

The IST program document and the request for the alternative or exemption would typically describe the method for selecting the interval dates, specifying the dates at which the interval will begin and end, and comparing the effect of those dates with that of the dates that would otherwise be required.

Basis for Recommendation

The staff believes that conducting IST programs for multiple unit sites using same Code edition could provide an improvement in program effectiveness.

3.3.3   Implementation of Updated Programs

Updating the IST program to a revised edition and addenda of the Code is an extensive effort that involves changes to administrative and implementing procedures. Often, the revised requirements will necessitate establishing new reference values, such as by implementing a vibration program using velocity measurements rather than displacement measurements for pump testing. Implementing a new comprehensive pump test may be necessary for parameters that are not currently measured. New "reference values" for currently monitored parameters may not be necessary if previous reference values were acceptable. However, the Code does not specifically require licensees to establish new reference values simply because a later edition is used.

NRC Recommendation

The NRC staff recommends that, before beginning the first tests during the new interval, licensees revise the implementing procedures according to the appropriate requirements. When the testing requires baseline values to be reestablished to meet Code changes, this would typically involve establishing the new baseline (reference) values during the first quarterly or cold shutdown outage test performed in the new interval, if not before. Before performing tests during the first refueling outage, licensees would typically revise implementing procedures for the tests to be performed during that outage to incorporate any new requirements or components.

Before or during startup from the refueling outage, licensees must complete all tests that are required to be performed during the refueling outage, as required by the Code (ISTC-3500 and Appendix I, Section I-1300). If a licensee determines that timely implementation is not possible, the staff recommends that the licensee submit a schedule to the NRC (1) before the beginning of the interval, or (2) before startup from the refueling outage if the interval begins while a plant is shut down for refueling.

For 120-month updated programs, the staff recommends that licensees submit relief requests before the next inspection interval's start date to allow adequate time for NRC review and approval within 12 months after the interval start date (i.e., submit the updated program at least 3 - 6 months before the start date.)

In accordance with the regulations, when updating a program to a later edition of the ASME Code, licensees must implement the updated program at the beginning of a 120-month interval. The regulations state that, where a pump or valve test specified by the Code is determined to be impractical and is not included in the revised IST program interval such that the basis for this determination must be submitted for NRC review and approval not later than 12 months after the expiration of the initial 120-month interval of operation from the start of the facility commercial operation and each subsequent 120-month interval of operation during which the test is determined to be impractical. However, experience has shown that licensees also identify impractical test provisions throughout the interval. In such cases, the staff recommends that licensees request relief as soon as they identify the condition. Because the requirements are impractical, the licensee would test the applicable components using the method proposed in the relief request in the period of time from the beginning of the new interval (or from the time of identification) until the NRC staff completes its evaluation. This would include for example, a situation in which a licensee identifies a solenoid valve that is within the scope of the IST program and is stroke-time tested but has no position indication, or if the licensee cannot meet the Code requirements because of design limitations and an alternative method may not comply with the Code requirements). Proposed alternatives to the Code requirements (rather than relief from "impractical" requirements) shall not be implemented until the NRC staff completes its evaluation (e.g., if a licensee proposes to implement a pump vibration program based on using spectral analysis, rather than the Code-specified method, the licensee must continue to meet the Code requirements until the NRC staff completes its evaluation).

Basis for Recommendation

When licensees update their IST programs to a revised edition and addenda of the Code, the NRC staff recognizes that changes might be completed over a period of time to allow for adequate review and approval; however, the staff recommends completing the procedural revisions in a timely manner. The regulations do not allow a licensee to continue with a previous program while waiting for NRC review and approval of the relief requests and proposed alternatives for the next interval. The staff recommendation that the request be submitted 3-6 months before the end of the inspection interval is based upon the expected time needed for the staff to evaluate the request and advise the licensee.

3.3.4 General Comments on Inservice Testing Intervals

The NRC has received requests for IST programs and partial submittals that lack the dates of the intervals or the Code edition in use. Some licensees were not aware that NRC may issue final rules amending 10 CFR 50.55a which are not reflected in the current printed version of the *Code of Federal Regulations*. Therefore, when those individuals revised their programs, they used the printed version of 10 CFR Part 50, "Domestic Licensing of Production and Utilization Facilities," to determine the Code edition cited in 10 CFR 50.55a(b) 12 months before the interval start date. However, a more recent edition may have been incorporated by reference in 10 CFR 50.55a(b) as noticed in the *Federal Register*, which may have resulted in the program being developed to an incorrect edition of the Code.

Additionally, several licensees have asked questions concerning phasing in the updated program. Generally, this is an acceptable approach for testing if the program does not involve any requests for relief from Code requirements.

NRC Recommendation

The NRC staff recommends that licensees include the interval dates and Code edition in each IST submittal, regardless of whether it is for an entire program or only a partial submittal containing new or revised relief requests. The staff must ensure that the interval dates are correct and that the evaluation is performed using the edition of the Code from which the licensee is requesting relief. The staff also recommends that licensees implement procedures to ensure that the individuals responsible for developing and maintaining the IST program are made aware of the regulatory changes made in 10 CFR 50.55a throughout the year.

For phasing in a new edition or addenda of the Code before the start of a new interval (or during an ongoing interval), the staff recommends that licensees submit a plan and schedule to the NRC. If there are no issues that require NRC review, the testing can be phased into the appropriate edition of the Code (1) during the 12 months prior to the interval start date, or (2) during any time period identified by the licensee up to an interval start date, if the phasing-in begins in the middle of an interval and a licensee wants to use an edition of the Code that is more recent than that incorporated by reference in 10 CFR 50.55a(b).

Basis for Recommendation

The NRC staff has noted incorrect interval dates and Code editions cited in IST program submittals. The Code stipulates that licensees shall calculate the IST interval according to the number of calendar years that have passed since the power unit was placed into commercial service. Licensees may verify the licensing and commercial operation dates for their plants by reviewing the annual "NRC Information Digest" (NUREG-1350). For convenience, the licensees for several plants have established the initial interval as beginning on the date of their operating licenses or some other unspecified milestone. However, the staff cautions that if the NRC revised 10 CFR 50.55a after the interval start date cited by the licensee and before the date of the operating license, and if the revision of 10 CFR 50.55a incorporated a later edition of the Code, the regulations may require use of a more recent edition than the licensee believes is required. Therefore, it is important that the IST program document state the Code edition and addenda used to develop the program, so that the NRC may verify the licensee's correct use of the applicable Code edition and addenda.

3.4   Skid-Mounted Components and Component Subassemblies

The Code class piping systems at a plant may include skid-mounted components or component subassemblies, such as valves in diesel air-start subassemblies, diesel skid-mounted fuel oil pumps and valves, steam admission and trip throttle valves for HPCI or auxiliary feedwater (AFW) pump turbine drivers, steam traps, and air supply system check valves and solenoid-operated valves for main steam isolation valves. If the licensee's safety analysis report (SAR) identifies these components as ASME Code Class 1, 2, or 3, they are subject to IST required by 10 CFR 50.55a. By contrast, if the SAR does not identify these components as ASME Code Class 1, 2, or 3 (or indicates that they are maintained as Code class, but are not required to be Code class), they are not subject to IST in accordance with 10 CFR 50.55a. Nonetheless, these components may be subject to periodic testing in accordance with Appendix A, "General Design Criteria for Nuclear Power Plants," and Appendix B, "Quality Assurance Criteria for Nuclear Power Plants and Fuel Reprocessing Plants," to 10 CFR Part 50.

NRC Recommendation

Subsections ISTB-1200(c) and ISTC-1200(c) define the components that are subject to IST. The staff has determined that testing the major component is an acceptable means to verify the operational readiness of the skid-mounted components and component subassemblies if the licensee discusses this approach in the IST program document. Licensees should consider and document the specific measurements and attributes of major component testing which relate to the assessment of skid-mounted component condition. In addition, various continuous and periodic observations of the major components (such as System Monitoring Walkdowns or Operator Logs) may also support assurance of skid-mounted component readiness. This is acceptable for both Code class components and non-Code class components that are tested and tracked by the IST program.

Basis for Recommendation

Various pumps and valves that are procured as part of larger component subassemblies are often not designed to meet the requirements for components in ASME Code Classes 1, 2, and 3. In RG 1.26, "Quality Group Classifications and Standards for Water-, Steam-, and Radioactive-Waste-Containing Components of Nuclear Power Plants," the NRC gives guidance on classifying components for quality groups A, B, C, and D (Code Classes 1, 2, and 3, and ASME VIII/ANSI B31.1, "Power Piping," respectively). (For additional guidance, licensees should review Section 3.9.6 of NUREG-0800, the NRC's Standard Review Plan.) When many of the components were procured, the requirements for IST did not apply and, thus, the components may not have included features for IST. Licensees may, therefore, elect to use the IST program for testing these components and state in the IST program document that the surveillance tests of these components adequately test the skid-mounted components.

The OM Code addresses both components that are physically mounted on the skid, and those that are not mounted on the skid but function the same as skid-mounted components (e.g., check valves in the service water system that supply cooling water to a pump), provided that testing the major component is adequate to test the function of the system component.

For components that are outside the scope of 10 CFR 50.55a, relief requests are not necessary. The NRC's position concerning testing components that are outside the scope of 10 CFR 50.55a is discussed in Section 2.2.3. Testing of skid-mounted check valves are specifically discussed in Section 4.1.10.

3.5   Pre-Conditioning of Pumps and Valves

3.5.1   Background

The regulations in 10 CFR 50.55a require licensees to test pumps and valves at nuclear power plants to assess their operational readiness within the scope of the ASME OM Code. Criterion XI, "Test Control," in Appendix B to 10 CFR Part 50 specifies that licensees must establish a test program to ensure that all testing required to demonstrate that structures, systems, and components (SSCs) will perform satisfactorily in service is identified andperformed in accordance with written test procedures that incorporate the requirements and acceptance criteria contained in applicable design documents. Criterion XI further requires that test procedures must include provisions to ensure that (1) all prerequisites for the given test have been met, (2) adequate test instrumentation is available and used, and (3) the test is performed under suitable environmental conditions. Criterion XI then requires licensees to document and evaluate the test results to ensure that the test requirements have been satisfied. In order to effectively assess operational readiness, the performance of the tested pump or valve, and the conditions under which the pump or valve must be capable of performing its safety function, need to be fully understood. Any maintenance activities performed before actual inservice testing are called "preconditioning" or "grooming" and this will adversely affect the validity of the test results.

3.5.2 NRC Guidance

In Information Notice (IN) 97-16, "Preconditioning of Plant Structures, Systems, and Components Before ASME Code Inservice Testing or Technical Specification Surveillance Testing," the NRC staff discussed the longstanding concern regarding unacceptable preconditioning of plant SSCs before testing. The staff noted that experience has demonstrated that some testing cannot be performed without disturbing or altering the equipment. The staff also stated that any such disturbance or alteration would be expected to be limited to the minimum necessary to perform the test and to prevent damage to the equipment. In addition, the staff alerted licensees that, in certain cases, the safety benefit of some preconditioning activities might outweigh the benefits of testing in the as-found condition.

The staff has provided guidance to the NRC's regional offices and inspectors with respect to preconditioning of plant equipment prior to ASME Code and TS testing. This guidance is found in the following documents (see Section 9 of Appendix A for the locations of these documents):

- NRC memorandum, dated July 2, 1996, from Frederick J. Hebdon, Director, Project Directorate II-3, Division of Reactor Projects I/II, Office of Nuclear Reactor Regulation, to Jon R. Johnson, Acting Director, Division of Reactor Projects, Region II, in response to

Technical Assistance Request TIA 96-007: "Regulatory Acceptability of Lubricating Valves Prior to Surveillance Testing"

- NRC Inspection Manual, Part 9900, "Technical Guidance: Maintenance - Preconditioning of Structures, Systems, and Components Before Determining Operability"

- Attachment 22, "Surveillance Testing," to IP 71111, "Reactor Safety: Initiating Events, Mitigating Systems, Barrier Integrity"

The guidance provided in these documents is instructive for nuclear plant personnel in providing assurance that testing conducted as part of the IST program is capable of assessing the operational readiness of pumps and valves within the scope of the ASME OM Code. NRC Inspection Manual, Part 9900, defines preconditioning as the "alteration, variation, manipulation, or adjustment of the physical condition of an SSC before Technical Specification surveillance or ASME Code testing." The licensee may consider several factors in determining whether an activity constitutes acceptable preconditioning of a pump or valve to be tested. For example, an activity would constitute acceptable preconditioning of a pump or valve if it is performed to protect personnel or equipment, or to meet the manufacturer's recommendations (or based on industry-wide operating experience). If a preventive maintenance activity (such as valve stem lubrication or pump venting) periodically occurs prior to testing, the licensee might justify the acceptability of this infrequent preconditioning of a pump or valve if the licensee evaluates the effect of the activity on the overall ability to assess the operational readiness of the pump or valve, and to trend degradation in its performance. As noted in the inspection guidance, the licensee should have evaluated and documented the activity as acceptable preconditioning before performing the testing.

Some activities would constitute unacceptable preconditioning of a pump or valve to be tested under the IST program. NRC Inspection Manual, Part 9900, defines unacceptable preconditioning of an SSC as an activity that alters one or more of the SSCs operational parameters and, thereby, results in acceptable test results. For example, a preventive maintenance activity might constitute unacceptable preconditioning of a pump or valve if the licensee routinely conducts the activity prior to testing. NRC Inspection Manual, IP 71111, Attachment 22, instructs NRC inspectors to evaluate the acceptability of any preconditioning of equipment in preparation for surveillance tests. Similarly, NRC inspectors also verify that licensees do not routinely schedule preventive maintenance activities prior to testing in order to help ensure that the test is passed satisfactorily. In addition to activities related to an individual pump or valve, maintenance or surveillance activities involving several SSCs, including the scheduling or timing of such activities, can inadvertently result in unacceptable preconditioning of a pump or valve.

NRC Inspection Manual, Part 9900, "Technical Guidance," provides a series of questions that NRC inspectors should consider when evaluating the acceptability of an activity that appears to involve preconditioning of a plant SSC. With respect to pumps and valves, those questions can be interpreted as follows:

- Does the practice performed ensure that the pump or valve will meet its testing acceptance criteria?

- Would the pump or valve have failed the test without the preconditioning?

- Does the practice bypass or mask the as-found condition of the pump or valve?

- Is preventive maintenance routinely performed on the pump or valve just before testing?

- Is preventive maintenance on the pump or valve performed only for scheduling convenience?

According to NRC Inspection Manual, Part 9900, "Technical Guidance," an activity constitutes unacceptable preconditioning if an affirmative answer is determined in response to any of these questions, and the activity meets the definition of unacceptable preconditioning provided in the inspection guidance. Licensees are encouraged to consider such questions as part of their determination of whether an activity related to a pump or valve in their IST program constitutes unacceptable preconditioning.

3.5.3 ASME Code Guidance

The ASME Code relies on the licensee to determine whether an activity would constitute acceptable or unacceptable preconditioning of a pump or valve prior to testing under its IST program, except in a few limited instances. One such instance is found in Sections I-3300 and I-7300 of the OM Code's mandatory Appendix I, "Inservice Testing of Pressure Relief Devices in Light-Water Reactor Nuclear Power Plants," which specify that no maintenance, adjustment, disassembly, or other activity that could affect as-found set-pressure or seat tightness data for pressure relief devices is permitted before testing. Another instance is found in Section 3.3 of ASME Code Case OMN-1, "Alternative Rules for Preservice and Inservice Testing of Certain Electric Motor-Operated Valve Assemblies in Light-Water Reactor Power Plants, OM Code-1995, Subsection ISTC," which specifies that inservice tests of motor-operated valves shall be conducted in the as-found condition, and that maintenance activities shall not be conducted if they might invalidate the as-found condition for inservice testing. The 2009 Edition of the ASME OM incorporates Code Case OMN-1 (including the as-found testing provision), and OMN-11, "Risk Informed Testing of Motor-Operated Valves," as Appendix III to the OM Code, to replace quarterly MOV stroke-time testing with periodic exercising and diagnostic testing. (Note: The details related to the ASME OM-2009 are for information only.) Where the ASME Code does not specify provisions for as-found testing of a pump or valve, the licensee is responsible for determining whether an activity constitutes acceptable or unacceptable preconditioning of a pump or valve to be tested under its IST program.

3.5.4 NRC Recommendation

The NRC staff has provided examples of acceptable and unacceptable preconditioning of plant components prior to testing in such documents as IN 97-16 and NRC Inspection Manual, Part 9900. Where the ASME Code does not provide specific provisions related to as-found testing of a pump or valve in the IST program, the staff considers acceptable preconditioning to include such activities as (1) periodic venting of pumps, which is not routinely scheduled directly prior to testing but may occasionally be performed before testing; (2) pump venting directly prior to testing, provided that the venting operation has proper controls with a technical evaluation to establish that the amount of gas vented would not adversely affect pump operation;

(3) occasional lubrication of a valve stem prior to testing of the valve, where stem lubrication is not typically performed prior to testing; (4) unavoidable movement attributable to the setup and connection of test equipment; and (5) test instrument venting directly prior to equipment testing. In each instance of acceptable preconditioning, the staff will expect the licensee to have available a documented evaluation of the preconditioning activity and a justification for continued confidence in the capability of the IST program to assess the operational readiness of the pump or valve. Generic evaluations may be acceptable as long as the evaluation bounds the conditions of specific activity performed on the SSC.

By contrast, the staff considers unacceptable preconditioning of pumps and valves in the IST program to include such activities as (1) routine lubrication of a valve stem prior to testing the valve; (2) operation of a pump or valve shortly before a test, if such operation could be avoided through plant procedures with personnel and plant safety maintained; and (3) venting a pump immediately prior to testing without proper controls and scheduling. Licensees may evaluate applicable NRC staff documents to determine whether specific activities prior to testing constitute acceptable or unacceptable preconditioning of a pump or valve in the IST program. The NRC staff encourages licensees to contact their NRC resident inspector or Office of Nuclear Reactor Regulation (NRR) or Office of New Reactors (NRO) project manager if questions arise regarding potential preconditioning of a pump or valve to be tested under the IST program.

3.6   Testing in the As-Found Condition

The intent of IST is primarily to detect and monitor the degradation or rate of changes of a component after a period of operation, or stand-by conditions. NRC staff review of IST programs during 1985 through 1991, Section XI, IWP and IWV, and Part 6 and Part 10 of OM-1987 had been interpreted to require that IST be performed in the "as-found" condition. However, later OM Codes, with the exception of Appendix I for safety/relief valves and Appendix III for MOVs, do not specifically require licensees to test components in the "as-found" condition. If "as-found" tests were not performed, degradation mechanisms or rate of degradation resulting from previous period of operation or stand-by condition can not be identified and obtained. Therefore, this section should ensure that the intent of IST be retained.

The Code does not specifically require licensees to test components in the "as-found" condition (except for safety and relief valves and, in the 2009 Edition of the ASME OM, MOVs) (Note: The details related to the ASME OM-2009 are for information only). Sections 1300, 1350 and 3300 of the ASME OM Code's mandatory Appendix I require licensees to measure the initial lift of safety relief valves to determine whether additional valves are to be tested. In addition, Sections 3300 and 7300 of Appendix I specify that licensees must periodically test all pressure relief devices and may not perform any maintenance, adjustment, disassembly, or other activity that could affect the as-found set pressure or seat tightness data before testing.

The "as-found" condition is generally considered to be the condition of a valve without pre-stroking or maintenance. The ASME OM Code does not require stroke-time testing or check valve stroking prior to maintenance; however, degradation mechanisms may not be identified if the licensee does not perform any as-found testing. However, the staff encourages licensees to perform as-found testing, where practical. The staff also cautions licensees to consider the timing of maintenance with regard to the required test intervals and the potential for pre-conditioning. Post-maintenance testing is required when the maintenance could have affected the valve's performance As-found testing may also apply to pumps in a similar fashion. Most inservice testing is performed in a manner that generally represents the condition of a standby component if it were actuated in the event of an accident (i.e., no pre-conditioning prior to actuation).

3.7   Testing at Power

In an effort to shorten refueling outages, many licensees are trying to perform as much maintenance, testing, and surveillance as possible with the nuclear power plant on line. For example, several licensees have submitted alternative requests to obtain NRC approval to conduct inservice testing once per refueling cycle, rather than during the refueling outage as prescribed by the Code. In preparing (and evaluating) such alternative requests, licensees (and the NRC staff) should consider several factors to ensure that the licensee's proposed alternative provides an acceptable level of quality and safety.If a licensee is testing a particular valve during refueling outages, the licensee may have determined that it is impractical to test the valve quarterly during operation. The licensee's IST program document should, therefore, discuss the basis for deferring the testing from quarterly (and during cold shutdowns) to refueling outages. Relief requests to perform testing once each refueling cycle with the nuclear power plant on line should be prepared in lieu of the refueling outage justification for each affected valve or group of valves. If necessary, the licensee should revise the refueling outage justification to be consistent with the relief request.

Licensees (and the NRC staff) should also consider whether the testing can be accomplished within the allowed outage time permitted by any applicable technical specification. In general, the time necessary to complete the testing should be significantly less than the allowed outage time. This is to preclude TS violations or the need to issue exigent TS amendments or notices of enforcement discretion (NOEDs). In addition, licensees should not conduct non-corrective maintenance/testing activities at power if the associated post-maintenance testing cannot reasonably be accomplished until the next outage.

Sometimes, there is a tradeoff between testing components at power (e.g., when they could be needed to mitigate the consequences of an accident) and testing them during outages (e.g., when there may be greater reliance on shutdown cooling or when other equipment is necessarily out-of-service). Licensees should quantitatively or qualitatively address the risks associated with testing components on line, rather than testing during the refueling outage. If the proposed testing could have a significant risk impact, or if its justification includes risk-related arguments, the relief request should be prepared and reviewed in accordance with RG 1.174 and Appendix D to Standard Review Plan (SRP) Chapter 19, as applicable. Licensees should also identify any compensatory measures to be established as a means to reduce the impact (e.g., risk and operational worker safety) of testing with the nuclear power plant at power.[1]

If relevant, licensees should also provide information on how testing at power (rather than testing during refueling outages) will affect scheduled maintenance work windows for the applicable system (i.e., whether the testing can be completed within the work windows or whether it will extend either the shutdown or at-power work windows). In addition, licensees will need to develop a new estimate of the maintenance unavailabilities that reflects the increased maintenance activities at power, and will need to document the basis for the new estimate (e.g., use plant logs or maintenance data to include in the current estimate of the maintenance unavailabilities those activities that were being performed during shutdown that will now be performed at power) [1].

At times, testing (or the disassembly and inspection of components) during refueling outages can be more advantageous than at-power operations from a worker safety perspective (for example, systems may be cold and depressurized). When requesting NRC authorization to perform testing with the nuclear power plant on line, licensees should consider worker safety and should discuss whether the applicable components can be adequately isolated and restored.

In Section 11.2.3 of Nuclear Management and Resources Council (NUMARC) 93-01, Revision 2, "Industry Guideline for Monitoring the Effectiveness of Maintenance at Nuclear Power Plants," the Nuclear Management and Resources Council. NUMARC, now NEI, provided additional guidance for conducting online maintenance and testing. It states, in part –

> Online maintenance [and testing] should be carefully managed to achieve a balance between the benefits and potential impacts on safety, reliability or availability. For example, the margin of safety could be adversely impacted if maintenance is performed on multiple equipment or systems simultaneously without proper consideration of risk, or if operators are not fully cognizant of the limitations placed on the plant due to out of service equipment. Online maintenance should be carefully evaluated, planned and executed to avoid undesirable conditions or transients, and to thereby ensure a conservative margin of core safety.

3.8 Potential Adverse Impact on Plant Components from Flow-Induced Vibration

Nuclear power plant operating experience has revealed that plant components can be adversely affected by flow-induced vibration caused by hydrodynamic and acoustic loading in nuclear power plant fluid systems. For example, BWRs have experienced damage to plant components in main steam systems as a result of vibration caused by acoustic resonance during power uprate operation. Flow-induced vibration has also resulted in damage to components in the reactor coolant system at a PWR plant. Licensees of new nuclear power plants, and licensees of currently operating plants planning to implement a power uprate, should ensure that the potential for adverse effects on plant components from flow-induced vibration is addressed during plant startup or power uprate ascension.

3.9 Repair of Pumps and Valves for New Reactors Under Construction

In a final rule issued on August 28, 2007 (72 FR 49352), the NRC amended its regulations to clarify the applicability of various requirements to each of the licensing processes by amending 10 CFR 50.55a in which the NRC clarified how the regulations in 10 CFR 50.55a apply to

---

[1] It should be noted that the assessment of risk resulting from performance activities as required by 10 CFR 50.65(a)(4) of the Maintenance Rule is not sufficient justification for testing components at power. This assessment is required for maintenance activities performed during power operations or during shutdowns. This configuration risk management does not address the relative merits of testing at power verses testing during refueling outages.

approvals, certifications, and licenses issued under 10 CFR Part 52. The NRC modified 10 CFR 50.55a to state that each combined license for a utilization facility is subject to the conditions in 10 CFR 50.55a, but is only subject to the conditions in 10 CFR 50.55a(f) and (g) after the NRC makes the finding under 10 CFR 52.103(g). This 52.103(g) finding is commonly associated with the date of initial fuel loading. Essentially, the effect of these changes to 10 CFR 50.55a is that combined license plants are subject to the requirements of Section III of the ASME *Boiler and Pressure Vessel Code* (ASME B&PV Code) from the date of issuance of the combined license during which the plant is under construction, and are subject to the requirements of Section XI of the ASME B&PV Code only after initial fuel loading. As a result of these changes to 10 CFR 50.55a, the question, "Which repair requirements (i.e., those under Section III or XI) apply to a combined license plant under construction?" was identified by industry and NRC staff in public meetings on ITAAC maintenance. In operating plants, repairs to ASME Code components are made using the rules of Section XI of the ASME Code; not Section III. The industry questioned whether the requirements of Section XI may be used when making repairs to an installed component in a plant under construction after the system or component has received its Code Symbol Stamp, but before the Commission makes its finding under 10 CFR 52.103(g). As a result, the NRC staff brought these questions to the ASME B&PV Code committees for their interpretation and possible action.

In 2010, the ASME Boiler Code committees examined the history and background of this issue and found that the ASME B&PV Code permits repairs using either Section III or XI. However, the ASME B&PV Code did not provide clear requirements on how repairs are to be accomplished under Section III. As a result, the Code committees developed definitive rules clarifying how repairs are to be performed under Section III for a plant under construction. Although the initiative was originally intended for application to new reactors, a similar and more urgent concern also existed for the Watts Bar Nuclear Plant, Unit 2 which recently restarted its construction activities and committed to use Section III for its plant's repairs during construction. These definitive rules were issued in ASME Code Cases N-801 and N-802 to address the concerns at Watts Bar, Unit 2. ASME Code Case N-802 clarifies how repairs shall be made at the plant site by the organization that originally manufactured or certified the component, and ASME Code Case N-801 clarifies how repairs shall be made at the plant site by organizations other than the original component manufacturer. ASME Code Case N-801 was needed at Watts Bar, Unit 2 because many manufacturers of components installed in the Watts Bar plant are no longer in the nuclear business. ASME Code Case N-801 is equally important for new reactors in which major plant components may be manufactured and certified in other countries, and for which it might be impracticable for those original foreign manufacturers to perform repairs once the component is installed in a U.S. plant. The NRC, therefore, finds Code Cases N-801 and N-802 to be applicable and appropriate for use by new reactors licensed under 10 CFR Part 52.

The NRC is currently preparing an amended rule in 10 CFR 50.55a permitting the use of ASME Code Cases N-801 and N-802 by new reactors under construction and licensed under 10 CFR Part 52.

4. SUPPLEMENTAL GUIDANCE
ON INSERVICE TESTING OF VALVES

The NRC staff has developed the following recommendations for valves that may be a part of an IST program at nuclear power plants. The types of valves discussed herein are covered by the ASME OM Code. Specifically, these include check valves (Section 4.1), power-operated valves (POVs) (Section 4.2), safety/relief valves (S/RVs) (Section 4.3), and other miscellaneous valves (Section 4.4).

4.1   Check Valves

The NRC staff considers check valves, and other automatic valves designed to close without operator action following an accident considered to be "active" valves that would be classified as such in the IST program (reference, for example, Section B 3.6.3 of the NUREG-1431, Revision 4, Westinghouse Standard Technical Specifications). Similar criteria could be applied to the opening function of a check valve. The flow through a check valve would be blocked by any condition that precludes flow through the system. For example, installing a flange or closing another valve (other than a check valve) in the line would block flow. A valve that is "positively held in place" would be one that has an operator or other auxiliary device that maintains the disk in an open or closed position, such as a stop check valve. SECY-77-439, "Single Failure Criterion," dated August 17, 1977, which was referenced in several plants' licensing bases, discusses the failure of a check valve to move to its correct position as a passive failure; however, this does not correspond to the issue of "active" versus "passive" for the purpose of IST.

The ASME OM Code defines valves that are self-actuating in response to some system characteristic, such as flow direction, for fulfillment of the required function(s) as Category C valves. The Code also defines valves for which seat leakage is limited to a specific maximum amount in the closed position for fulfillment of their required function(s) as Category A valves. Those check valves (Category C valves) that must also be leak-tight (Category A valves) would be designated as Category A/C in the IST program.

Whereas the OM Code only requires licensees to exercise Category C check valves on a periodic basis, Category A/C check valves must be leak tested in addition to being exercised. The NRC staff has found that, in many instances, licensees are not assigning check valves to Category A/C, despite the fact that the licensees take credit for the check valve providing an essentially leak-tight function.

When determining the proper categorization of a check valve, a licensee should consider all applicable aspects. For example, the licensee should determine (1) whether the flow requirements for connected systems can be achieved with the maximum possible leakage through the check valve, (2) the effect of any reduced system flows resulting from the leakage on the performance of other systems and components, (3) the consequences of the loss of water from the system, (4) the effect that backflow through the valve may have on piping and components, such as the effect of high temperature and thermal stresses, and (5) the radiological exposure to plant personnel and the public caused by the leak. If any of the above considerations indicate that Category C testing may not be adequate, licensees should assign the check valve to Category A/C and should comply with the associated leak testing requirements.

Licensees may refer to NRC Information Notice (IN) 91-56, "Potential Radioactive Leakage to Tank Vented to Atmosphere," dated September 19, 1991, for information on the categories assigned to valves that function to close. These valves may also function to prevent leakage above an assumed limit to prevent the plant from exceeding the limits in 10 CFR Part 100, "Reactor Site Criteria." Section 4.1.1 herein discusses backflow testing of check valves in series.

Subsection ISTC-3550 of the OM Code discusses valves in regular use and states that valves that operate in the course of plant operation at a frequency that would satisfy the exercising requirements need not be additionally exercised, provided that the observations otherwise required for testing are made and analyzed during such operation, and recorded in the plant record at intervals no greater than specified in Subsection ISTC-3510. Even if licensees "exercise" check valves in accordance with Subsection ISTC-3550, they need to be included in the valve list in the IST program document, and the record (e.g., plant log, test procedure) needs to indicate that the test requirements are met.

For grouping valves in multiple nuclear power plants of like design and construction, if the plants are "identical" and the grouped valves have similar operational experience and otherwise meet the grouping criteria, it is acceptable for licensees to group valves from multiple plants. If a potentially generic problem is identified through disassembly and inspection during a refueling outage, the licensee must inspect all valves in the group in that plant during the refueling outage. If the other plant is not also in a refueling outage, inspection of the valves in the group that are installed in that plant may be deferred to the next refueling outage unless the licensee's evaluation of the problem indicates that it could impact the safety of continued operation. "Grouping" may also be applied to the use of nonintrusive techniques as discussed in Section 4.1.2 (below), although the focus is slightly different, in that all of the valves in the group are tested, while the nonintrusive techniques are applied to only one valve of the group; therefore, all valves in the group must be in the same nuclear power plant.

The NRC issued the following Information Notices and Bulletins (BL) on IST for check valves:

- IN 82-08 "Check Valve Failures on Diesel Generator Engine Cooling System"
- IN 83-54 "Common Mode Failure of Main Steam Isolation Non-Return Check Valves"
- IN 88-70 "Check Valve Inservice Testing Program Deficiencies"
- IN 2000-21 "Detached Check Valve Disk Not Detected by Use of Acoustic and Magnetic Nonintrusive Test Technique"
- BL 83-03 "Check Valve Failures in Raw Water Cooling System of Diesel Generators

4.1.1 Closure Verification for Series Check Valves without Intermediate Test Connections

Some plants have piping configurations that include two check valves in series with no provision for verifying that each valve can close. These valves may perform a safety function in the closed position. For example, the valves may be required to prevent the gross diversion of flow rather than to be leak-tight. The Code requires that each valve performing safety functions must be stroked to the position(s) required for the valve to perform those functions. The requirements for testing two check valves in a series configuration are addressed in Subsection ISTC-5223, "Series Valve Pairs," as follows:

If two check valves are in a series configuration without provisions to verify individual reverse flow closure (e.g., keepfill pressurization valves) and the plant safety analysis assumes closure of either valve (but not both), the valve pair may be operationally tested closed as a unit. If the plant safety analysis assumes that a specific valve or both valves of the pair close to perform the safety function(s), the required valve(s) shall be tested to demonstrate individual valve closure.

NRC Recommendation

Both check valves in a series pair should be tested to demonstrate individual valve closure if the plant safety analysis credits or otherwise requires both valves. For example, General Design Criterion (GDC 14), "Reactor Coolant Pressure Boundary," requires licensees to test the valves in the reactor coolant pressure boundary to demonstrate extremely low probability of abnormal leakage. Pressure isolation valves are a special case of reactor coolant pressure boundary valves, which are generally required to be individually leakage tested at a frequency specified by the TS and the Code.

Systems containing series pair check valves may have provisions for verifying that at least one valve is capable of closing. These provisions enable the licensee to measure or observe operational parameters such as leakage, pressure, or flow during each quarter, each cold shutdown outage, or each refueling outage. However, testing the series pair as a unit provides no assurance that both valves close. The only indication of a problem would be the failure of both valves in the series. If the valve pair is operationally tested closed as a unit, as allowed by Subsection ISTC-5223, because the plant safety analysis assumes closure of either valve (but not both), and the tested unit closure capability is questionable, both valves should be declared inoperable and corrective actions must be taken for both valves.

If it is not practical to flow test the pair of check valves in accordance with the Code, the licensee may demonstrate the closure safety function of each valve by other positive means, such as nonintrusive testing, or disassembly and inspection. However, licensees should not use these methods to verify leak tightness, which requires Category A valve testing.

Basis for Recommendation

The requirements of Subsection ISTC-5223 are contained in the NRC's endorsement of the ASME OM Code 2004, Edition with 2005 and 2006 Addenda, as published in a regulatory amendment to 10 CFR 50.55a.

Subsections ISTC-5221(a) and/or ISTC-5221(c) allow the use of nonintrusive examination, disassembly and inspection, or other positive means, for check valves that have no provision for testing individual valves or practical means to demonstrate the closure capability of each valve by flow, but not for verifying leak-tightness.

Keep-fill valves are a special case, in that they are redundant check valves in a system in which only one valve of a series is actually necessary to perform a system's intended function. Licensees have proposed to exclude the upstream valve from the IST program. However, recognizing that neither valve can individually demonstrate a closure function, and that the Code alternative allows the valve pair to be operationally tested closed as a unit, the NRC staff's position is that both valves must be included in the IST program and should be operationally tested as a pair to prevent reverse flow. Upon observing leakage, the licensee should disassemble, inspect, and repair or replace both valves (as necessary) before return to service.

4.1.2 Exercising Check Valves with Flow and Nonintrusive Techniques

The Code requires both an open and close test for check valves and that licensees exercise check valves to the position(s) required to fulfill their safety function(s). To verify the disk position of check valves that do not have external disk position indication, the Code allows licensees to use indirect evidence (such as changes in system pressure, flow, temperature, or level) or other positive means. An acceptable test method must demonstrate by positive means that a check valve disk moves to the position necessary to fulfill its safety function. The demonstration may not require that the valve be exercised "full-open" to the backstop. The full-stroke to the open position may be verified either by passing design flow or by other positive means such as nonintrusive techniques. The "other positive means" must be repeatable to meet the intent of the Code.

NRC Recommendation

The licensee may use nonintrusive techniques for IST of check valves. Relief is not required except as would be necessary for the testing frequency if the test interval extends beyond each refueling outage as allowed by the OM Code. The licensee may use nonintrusive techniques to verify the valve's capability to open, close, and fully stroke if it is qualified for the application in accordance with the plant's quality assurance program requirements. A qualified nonintrusive technique is one that has been successfully and reliably demonstrated for the examination method and specific valve application. The licensee may qualify the technique and application on its own equipment, subcontract it to a vendor, or rely on the results of the Nuclear Industry Check Valve Group (NIC) evaluation of nonintrusive diagnostic techniques for check valves. Personnel training and qualification performed by a vendor in accordance with the licensee's quality assurance program may be acceptable; however, the technique must also be qualified as described above. Records of techniques and qualification documentation shall be maintained in accordance with the licensee's quality assurance program, and the NRC inspector may examine the licensee's records.

Basis for Recommendation

The NRC staff previously determined that nonintrusive testing methods appropriate for certain valve applications are acceptable to verify the capability of the valve to open, close, and fully stroke, provided that the licensee properly qualifies the testing methods used for the valve application in accordance with the plant's quality assurance program requirements. These techniques are considered "other positive means" in accordance with Subsection ISTC-5221(a)(3). It is the licensee's responsibility to qualify and document the results in accordance with the plant's quality assurance program requirements. Appendix B to 10 CFR Part 50 provides for the quality assurance program, which includes, in part, requirements to ensure that nondestructive testing is controlled and accomplished by qualified personnel using qualified procedures in accordance with applicable codes, standards, specifications, criteria, and other special requirements. The NIC conducted an experimental research and testing program to evaluate available nonintrusive technologies to determine their acceptability and reliability for use in check valve testing applications. Information on qualification of nonintrusive testing is provided in the summary of the NRC's public workshops on the revision of NRC Inspection Procedure 73756, dated July 27, 1995. In response to a question about expectations for qualification of a nonintrusive test method, the NRC indicated that a qualified nonintrusive test method is a technique that has been successfully and reliably demonstrated for the examination method and for the specific valve application. Other expectations discussed in NRC IN 2000-21 are summarized in the following paragraphs.

Qualification includes establishing a performance baseline when the check valve is known to be in good operating condition. A check valve's performance can be assessed against this baseline trace. One means to determine if a nonintrusive test method or technique will provide accurate, reliable, and repeatable results for a specific check valve is to qualify the method prior to its use. The qualification process may reveal that certain techniques or methods give inconclusive results for a particular application. Acoustic techniques and test methods are susceptible to other plant system noise being transmitted and masking or affecting the desired sound pattern and results. The NIC suggests the use of more than one technique to verify questionable results.

4.1.3   Full Flow Testing of Check Valves

The ASME OM Code requires both an open and close test for check valves and that licensees exercise check valves to the position(s) in which they perform their safety function(s). A check valve's full-stroke to the open position may be verified by passing the necessary flow through the valve to perform its safety function. This is considered an acceptable full-stroke. Any flow rate less than this is considered a partial-stroke exercise. A valid full-stroke exercise by flow requires that the flow through the valve must be known. Knowledge of only the total flow through multiple parallel lines does not provide verification of flow rates through an individual valve and may not be a valid full-stroke exercise without further analysis.

4.1.3.1   Alternative to Direct Flow Measurement

Flow through a check valve must be known for a valid full-stroke exercise test, but an alternative to direct flow rate instrumentation may be acceptable. Any quantitative measure that has acceptance criteria that demonstrate the required flow through the check valve may be used to satisfy the full-stroke requirement. An indirect measure of flow may be acceptable. For example, a change in tank level over a specified period could be used. In another case, the acceptance criterion could be based on a change in flow rate of an instrumented line when flow is admitted from a non-instrumented line containing the check valve being tested. In any event, some form of quantitative criteria should be established to demonstrate full-stroke capability.

4.1.3.2   Flow through Parallel Lines

Knowledge of total flow through multiple parallel lines does not provide indication of flow through each individual path. The objective of inservice testing is to evaluate and investigate the possibility of degradation of individual components and to take corrective action before a component fails. Verification of total header flow rate might not identify a problem, developing or occurring, with an individual check valve in one of the parallel flowpaths. With respect to the balancing of flow, TS requirements are based on the flow from one loop being lost through a postulated break. Consequently, that flowpath is restricted or throttled to minimize significant diversion of flow. TS surveillance requirements were not intended to verify individual check valve operability. The licensee is expected to justify the use of a test method that does not verify full-stroke of individual check valves.

For example, in a Beaver Valley Power Station, Unit 1, Safety Evaluation Related to the IST Program and Relief Requests dated January 24, 1992 (ADAMS Accession No. 9201310094), the NRC informed the licensee of the results of an evaluation of flow through parallel lines and stated that a flow test through parallel lines without individual flow measurement may not be sufficient to indicate that the check valves in the lines are full-stroke exercised. Knowledge of only total flow through multiple parallel lines does not provide verification of flow rates through an individual valve and may not be a valid full-stroke exercise without further analysis.

4.1.3.3   Accident Condition Flow

The phrase, "necessary flow through the valve for safety function," is the largest flow rate for which a licensee takes credit in a safety analysis for this check valve in any flow configuration. The safety analyses are those contained in the plant's final safety analysis report (FSAR), or equivalent, but are not limited to the accident and transient analyses.

4.1.3.4   Check Valves Not Required to Be Fully Opened

For check valves that are never required to open fully (i.e., thermal expansion or siphon breakers), verification of design (safety) function is testing to confirm the capability of forward flow through the system. In addition to verifying its safety function performance, licensees should develop quantifiable acceptance criteria for the testing of these components. Verifying that a system is full is an acceptable means for verifying that the keep-fill check valves are capable of opening to provide flow when necessary.

4-6

4.1.3.5 Alternative to Full-Flow Testing

Full-flow testing of a check valve may not be achievable for certain valves. It may be possible to qualify other techniques to confirm that a valve is exercised to the position required to perform its safety function. To substantiate the acceptability of any alternative technique for meeting the ASME Code requirements, licensees must, as a minimum, address and document the following items in the IST program. Any alternative techniques for meeting ASME Code requirements must be submitted to the NRC for authorization pursuant to 10 CFR 50.55a (a)(3)(i).

(1)   the difficulty of performing a full-flow test

(2)   a description of the alternative technique used and a summary of the procedures being followed

(3)   a description of the method and results of the program to qualify the alternative technique for meeting the ASME Code

(4)   a description of the instrumentation used and the maintenance and calibration of the instrumentation

(5)   a description of the basis used to verify that the baseline data has been generated when the valve is known to be in good working order, such as recent inspection and maintenance of the valve internals [components]

(6)   a description of the basis for the acceptance criteria for the alternative testing and a description of corrective actions to be taken if the acceptance criteria are not met.

The NRC staff's position with respect to full-flow testing of check valves allows licensees flexibility in qualifying alternatives to full-flow testing. In general, licensees should demonstrate that the alternative test is quantifiable and repeatable. The alternative test must meet the intent of the ASME OM Code. This qualification of the alternative test should be documented by the licensee and should be available for review by NRC inspectors. Any alternative techniques for meeting ASME Code requirements must be submitted to the NRC for authorization pursuant to 10 CFR 50.55a(a)(3)(i).

The 1994 Edition of the ASME OM Code and subsequent editions include the use of nonintrusive testing as other positive means for demonstrating check valve exercising. The criteria listed in the NRC staff's position for full-flow testing of check valves could be applied to the nonintrusive techniques.

4.1.4 Disassembly and Inspection Alternative to Flow Testing

Guidance regarding disassembly and inspection of certain check valves is included in Subsection ISTC-5221(c) for use if the test methods of Subsections ISTC-5221(a) and ISTC-5221(b) are impractical or sufficient flow cannot be achieved.

NRC Recommendation

Testing of check valves by disassembly should be accomplished as follows:

> The sample disassembly and inspection program should involve grouping similar valves and testing a different valve in each group during each refueling outage. The sampling technique should require that each valve in the group be the same design (manufacturer, size, model number, and materials of construction) and have the same service conditions including valve orientation.
>
> Additionally, at each refueling outage sampling the licensee should verify that the disassembled valve is capable of full-stroking and that the internals of the valve are structurally sound (no loose or corroded parts). Also, if the disassembly is to verify the full-stroke capability of the valve, the disk should be manually exercised. While the valve is in a partially disassembled condition the valve internals should be inspected and the condition of the moving parts evaluated. This inspection and evaluation should include verification that the valve disk is free to move. Following reassembly, a partial flow test should be performed, if practical.
>
> A different valve of each group should be disassembled, inspected, and manually full-stroke exercised at each successive refueling outage, until the entire group has been tested. If the disassembled valve is not capable of being full-stroke exercised or there is binding or failure of valve internals, then the remaining valves in that group should also be disassembled, inspected, and manually full-stroke exercised during the same outage. Once this is completed, the sequence of disassembly should be repeated unless the valve group is in the Condition Monitoring Program alternative and an extension of the interval can be justified.
>
> At least one valve from each group should be disassembled and examined at each refueling outage; all valves in each group shall be disassembled and examined at least once every 8 years.

The disassembly/inspection must be qualified to evaluate the condition of the valve and to assess its continued operability. The licensee is responsible for the development and implementation of a program to ensure that IST personnel are appropriately trained and qualified to perform the valve disassembly/inspections. Licensees should implement the provisions of ANSI/ASME N45.2.6, "Qualifications of Inspection, Examination, and Testing Personnel for Nuclear Power Plants." The staff encourages those licensees that review the ANSI standard for guidance in developing a program for the qualification of IST personnel.

Basis for Recommendation

In ASME OM Code, Subsection ISTC-5221(c) provides for the use of sample disassembly and inspection of check valves. Specifically, Subsection ISTC-5221(c) provides requirements for sample disassembly and inspection of certain check valves, if the test methods of Subsections

ISTC-5221(a) and ISTC-5221(b) are impractical, or if sufficient flow cannot be achieved or verified. ISTC-5221(c)(3) provides the requirements that every valve in each group must be examined at least once every 8 years. These Code requirements were incorporated by reference in 10 CFR 50.55a on September 22, 1999 (64FR51370).

4.1.5 Reverse Flow Testing of Check Valves

The OM Code requires that Category C check valves (those that are self-actuated in response to some system characteristic such as pressure or flow direction) that perform a safety function in the closed position to prevent reverse flow must be tested in a manner that proves that the disk travels to the seat on cessation or reversal of flow. In addition, for Category A/C check valves (those that have a specified leak rate limit and are self-actuated in response to a system characteristic), seat leakage must be limited to a specific maximum amount in the closed position for fulfillment of their function. Verification that a Category C valve is in the closed position can be achieved through visual observation, by an electrical signal initiated by a position-indicating device, by observation of appropriate pressure indication in the system, by leak testing, or by other OM Code-defined positive means, i.e. ISTC-5221.

ASME Code-class check valves that perform a safety function in the closed position and are frequently not reverse flow tested include the following examples:

(a)   main feedwater header check valves
(b)   pump discharge check valves on parallel pumps
(c)   keep-fill check valves
(d)   check valves in steam supply lines to turbine-driven AFW pumps
(e)   main steam non-return valves
(f)   chemical and volume control system (CVCS) volume control tank outlet check valves

4.1.5.1   Closure Capability of Check Valves that Do Not Have Defined Seat Leakage Limits

When performing a test to verify closure capability of a check valve that does not have a defined seat leakage limit, the achievement of the necessary system flow rate through the intended flowpath might be an adequate demonstration of the closure capability of a check valve. In addition, the licensee should evaluate the consequences of reverse flow through the check valve. This evaluation should consider the loss of water from that system and connecting systems, the effect that the leakage might have on components and piping downstream of the valve, and any increase in radiological exposure resulting from the leakage.

A plant's safety analysis may include a leakage limit for a particular valve, or may only require that the valve close to inhibit gross leakage. When a valve has a safety-related function to close to prevent diversion of flow between trains of a system, there may be a leakage limit based on the total system requirements. The Code does not specifically require these valves to be Category A. The basis for assigning valves to categories should be available for inspection.

Licensees may refer to IN 91-56, "Potential Radioactive Leakage to Tank Vented to Atmosphere," regarding information concerning the categories assigned to valves that function to close to prevent leakage above an assumed limit thus preventing a plant from exceeding the limits of 10 CFR Part 100.

4.1.5.2   Listed Systems

In reference to the examples of valves that are not frequently backflow tested (listed in Section 4.1.5, above), all of the listed systems do not necessarily apply to each plant. As a minimum, a licensee should evaluate the listed systems to determine whether they apply to its facility and should make any necessary modifications to its IST program. Examples (d), (e), and (f) are specific to PWRs, while Example (a) may apply to both PWRs and BWRs. Example (f) may serve an important safety function at some PWRs to separate the non-safety-grade water source from the safety-grade source.

4.1.5.3   Permissible Leak Rates

Subsection ISTC-3630(e) requires that leakage rate measurement shall be compared with the permissible leakage rate specified by the licensee for a specific valve or valve combination. If the licensee does not specify leak rates, permissible leak rates are provided in Subsection ISTC-3630(e). It should be noted that the OM Code does not provide either criteria or guidance concerning the methods licensees should use to establish or specify the permissible leak rate of a particular valve. The OM Code recognizes that the leak behavior of a valve varies according to valve type and size, vendor, service conditions, safety-related functions, and other factors, and there is no simple leak rate rule that applies to all valves.

In general, licensees should set the leak rate limits within certain bounds. If the leak limits are too low, unnecessary valve repairs or adjustments can result. Leak limits that are too high could result in failure of the tests required by Appendix J, "Primary Reactor Containment Leakage Testing for Water-Cooled Power Reactors," to 10 CFR Part 50, thereby leading to concerns regarding the leak-tight integrity of the containment. Appropriate permissible leak rates can only be developed and refined by analyzing and trending the leak rate data for specific valves or for similar valves at other plants. Therefore, the NRC staff has not provided specific guidance concerning leak rates. Licensees should document their methods for establishing the initial permissible leak rates and procedures for verifying compliance with the leak rate limits.

4.1.5.4   Closure Testing of Stop Check Valves

If a stop-check valve does not perform a safety-related function in the closed position, valve closure is only necessary to ensure a repeatable starting point for opening testing. Valves may be closed by using a handwheel or a hand switch. (Note: 1996 Addenda, and later Editions and Addenda to the OM Code require bidirectional testing.)

If the use of a handwheel or hand switch to close a valve achieves the safety-related function of the system, exercising the valve by this method meets the ASME Code requirements of Subsection ISTC-5221. By contrast, if closure of a stop check valve on cessation or reversal of flow is required to accomplish a safety-related function, its closure must be verified by reverse flow testing or other positive means, such as acoustic monitoring or radiography.

When no other means of verification are possible, licensees may disassemble valves to verify valve closure. However, disassembly provides limited information on valve capability to seat on cessation or reversal of flow. Furthermore, if the method involves extensive disassembly, a post-reassembly test would be necessary in accordance with Subsection ISTC-3310, because disassembly and inspection can increase the probability of human error when the valve is reassembled. Licensees may investigate the use of nonintrusive testing techniques and may implement such techniques if they are demonstrated to be effective to assess closure capability, degradation, and incipient failure. Infrequent disassembly and inspection of the valves are appropriate to assess overall check valve condition, while reverse flow testing and nonintrusive testing provide an assessment of continued operational readiness.

4.1.5.5   Other Positive Means of Verification

Subsection ISTC-5221 allows for other positive means of verification. Examples from IST programs include verifying that a parallel centrifugal pump does not spin in reverse to verify closure of a pump discharge check valve, monitoring an upstream pressure indicator, monitoring a tank level, measuring the flow rate of a redundant train, or opening an upstream vent and drain valve.

4.1.6   Extension of Test Interval to Refueling Outage for Check Valves Verified Closed by Leak Testing

When it is impracticable for the licensee to verify check valve closure or opening during plant operation or cold shutdown, it is acceptable for the licensee to extend the check valve quarterly exercise test (both open and close) to the refueling outage. The closure verification may be performed in conjunction with the Type C leak rate test conducted in accordance with Option A or Option B of Appendix J to 10 CFR Part 50. Licensees may also perform the open exercise test during the refueling outage or anytime during the fuel cycle interval.

NRC Recommendation

If no other practical means are available, it is acceptable for licensees to extend the quarterly closure or opening exercise test to a refueling frequency. In such instances, the licensee must develop a refueling outage justification describing the impracticality of performing the quarterly closure test during plant operation or cold shutdown. The NRC staff has determined that the need to set up test equipment constitutes adequate justification to defer reverse flow testing of a check valve to a refueling outage. By referencing the refueling justification in the IST program document, the licensee may perform the closure exercise test during each reactor shutdown for refueling. A seat leak test is one method to verify that the obturator has traveled to the seat. All requirements of each individual valve category are applicable, although repetition of a common testing requirement is not required. Therefore, when the required performance of the Appendix J leak rate test coincides with a refueling outage exercise seat leak test, only the Appendix J test is required.

The OM Code states that open and close tests need only be performed at an interval "when it is practicable to perform both tests." The OM Code also states that licensees are not required to perform open and close tests at the same time if they are both performed during the same interval. For example, since the closure test is extended to the refueling outage by the refueling justification, the quarterly exercising test may also be extended to the refueling outage or may be performed anytime during the fuel cycle interval.

Basis for Recommendation

OM Subsection ISTC-5221(a) states that valve obturator movement observations shall be made by a direct indicator (e.g., a position-indicating device) or other positive means including seat leak testing. Therefore, a seat leak test is one method to verify that the obturator has traveled to the seat.

OM Subsection ISTC-3522(c) states that if exercising is not practical during plant operation and cold shutdowns, it shall be performed during refueling outages. A refueling outage justification shall document the extension of the exercise test to the refueling outage.

OM Subsection ISTC-3522(a) states that open and close exercise tests need only be performed at an interval when it is practical to perform both tests. This Code section also states that open and close tests are not required to be performed at the same time if they are both performed during the same interval.

4.1.7 Testing and Examination of Check Valves Using Manual Mechanical Exercisers

ASME OM Code, Subsection ISTC-5221(b), "Valve Obturator Movement," in part, requires that if a manual mechanical exerciser is used to test the check valve, the force(s) or torque(s) required to move the obturator to fulfill its safety function(s) shall meet the acceptance criteria specified by the owner. This includes the following:

(1) Exercise test(s) shall detect a missing obturator, sticking (closed or open), binding (throughout obturator movement), and the loss or movement of any weights.

(2) Acceptance criteria shall consider the specific design, application, and historical performance.

(3) If it is impractical to detect a missing obturator or loss or movement of any weight(s), other positive means may be used (e.g., seat leakage tests and visual observations to detect obturator loss and the loss or movement of external weight(s), respectively).

NRC Recommendation

In the Code requirements through the OMa-1996 Addenda, Subsection ISTC 4.5.4(b) requires that, if a manual mechanical exerciser is used for IST movement of the obturator, the force or torque to initiate disk movement shall not vary by more than 50 percent from an established reference value. Licensees have continuously experienced difficulty with this IST acceptance criterion. The manual mechanical exerciser assembly includes a packing gland to seal the hinge pin penetration of the valve body. The hinge pin seal packing introduces conditions that produce variations in friction forces over time. These variations make it difficult to establish a reference value that would be continually consistent and appropriate for use in IST.

The NRC has received a number of requests for relief from the requirement that force or torque to initiate disk movement not vary by more than 50 percent of the reference value. Licensees have also requested that the ASME OM Working Group on Check Valves (WGCV) reexamine this requirement. The WGCV reexamination resulted in the Code change in OMa-2000, Subsection ISTC-5221(b), "Valve Obturator Movement," which is discussed above. This change requires the owner to specify the acceptance criteria within certain Code-defined expectations. In establishing an acceptance criterion for IST when using a manual mechanical exerciser, the owner should consider the interactions, wear and effects of the valve parts on friction forces, and the valve preventive maintenance activities.

Basis for Recommendation

Mechanical exercisers are attached to a hinge pin that is fixed to the disk and penetrates the valve body. Many of these valves involve swing check valves that manufacturers supplied with a lever arm and counterweight modification. The counterweight is used to affect the opening or closing response of the disk to flow conditions, depending upon the lever arm's location relative to the disk. The counterweight modification involves the use of a packing gland to seal the hinge pin penetration of the valve body. The seal packing introduces variations over time with regard to the required disk opening force and opening and closing responses of the disk, depending upon the type of packing material used, its condition, friction changes, leakage control adjustments, and the packing gland tightening procedure. Any wear of the hinge pin and bearing interfaces may exacerbate these variations. Disk opening and closing friction forces may also change as a result of valve preventive maintenance activities. ASME OM, Subsection ISTC-5221(b), "Valve Obturator Movement," states that if a mechanical exerciser is used to exercise the check valve, the force(s) or torque(s) required to move the obturator to fulfill its safety function(s) shall meet the acceptance criteria specified by the owner.

4.1.8 Check Valve Bidirectional Testing and Condition Monitoring Program

Bidirectional testing ensures that a check valve is adequately tested, regardless of its safety function. Such testing also improves the IST capability to detect valve degradation prior to valve failure. Two significant OM Code changes, in Subsections ISTC 4.5.4(a) and ISTC 4.5.5, respectively, were introduced in OMa-1996. Specifically, those changes included (1) a requirement for bidirectional exercise testing of the disk movement of check valves, and (2) a voluntary provision to use the condition monitoring program as an alternative to IST exercise testing for certain check valves. This integral two-part improvement to the Code provides interrelated requirements. The condition monitoring program allows licensees certain IST flexibility in establishing the types of test, examination, and preventive maintenance activities and their associated intervals, when justified based on the valve's performance and operating condition. These Code changes were developed so that licensees who elect not to implement condition monitoring in their IST program, would be required to bidirectionally test check valves as a default set of testing and examination requirements.

OMa-1996, Subsection ISTC 4.5.4(a), "Valve Obturator Movement," in part, requires that the necessary obturator movement during exercise testing shall be demonstrated by performing both an open and close test, and observations shall be made by a direct indicator (e.g., a positive-indicating device) or by other positive means (e.g., changes in system pressure, flow, level, temperature, seat leakage, testing, or nonintrusive testing and examination).

OMa-1996, Subsection ISTC 4.5.5, Condition Monitoring Program, in part, provides an option for the owner to establish a check valve condition monitoring program alternative to the testing and examination requirements of Subsections ISTC 4.5.1 - 4.5.4. The purpose of the program is to improve valve performance and optimize testing, examination, and preventive maintenance activities in order to maintain acceptable check valve performance. The program must be implemented in accordance with Appendix II, "Condition Monitoring Program."

NRC Recommendation

The required testing or examination of the check valve obturator movements to both the open and closed positions, as required by Subsection ISTC-3522, is necessary to assess the valve's operational condition, confirm the acceptability of its performance, and detect degradation prior to failure. Single-direction flow testing of check valves will not always detect functional degradation of the valves.

The NRC staff considers the condition monitoring program approach of ASME OM, Appendix II, for check valve IST with the regulatory modifications in 10 CFR 50.55a(b)(3)(iv), to be an improvement over present Code requirements, and encourages licensees to implement the condition monitoring alternative.

Basis for Recommendation

The NRC incorporated OMa-1996 by reference in 10 CFR 50.55a on September 22, 1999 (64FR51370). In OMa-1996, Subsection ISTC 4.5.4(a) included the requirement that the necessary obturator movement during exercise testing must be demonstrated by performing both open and closed tests. The NRC agrees with the need for a required demonstration of bidirectional exercise testing of the movement of the check valve disk. Single-direction flow testing will not always detect degradation of the valve. The classic example of the flawed single-direction testing strategy is that the loss of the disk would not be detected during forward flow tests. The detached disk could be lying at the bottom of the valve body or another part of the system, and could move to block flow or disable another valve or component. The most recent example of an undetected detached check valve disk lying at the bottom of the valve body is captured in the discussion of the event described in IN 2000-21. The NRC considers testing or examination of the check valve obturator movement to both the open and closed positions necessary to assess its condition and confirm acceptable valve performance.

The use of the ASME OM Code, Appendix II, "Condition Monitoring Program," as incorporated by reference in 10 CFR 50.55a on September 22, 1999 (64FR51370), includes requirements that apply when extending check valve IST intervals, with regard to consideration of the plant safety significance, justification by trending of current and historical valve condition and performance data, maximum IST interval limits, stepwise IST interval limits, and bidirectional testing or examination. These requirements provide the licensee with knowledge of the valve's operating condition, monitor and verify valve performance over extended intervals, and provide a process to justify prudent IST interval extensions to reduce the burden of unnecessary IST.

The ASME OM Code Committee, Working Group on Check Valves (WGCV) has completed proposed changes to Subsection ISTC of the ASME OM Code to address the regulatory modification issues. ASME issued these changes as part of OMb-2003. All changes discussed above are incorporated in the 2004 Edition of ASME OM Code.

4.1.9   Instrumentation Requirements

Instruments used to verify that check valves fulfill their safety function(s) are not subject to the same range and accuracy requirements as instrumentation used for pump-related IST. However, OM-1998 added Subsection ISTC-3800, "Instrumentation," to provide specific requirements for instrumentation that is used in the testing and examination of valves. Specifically, in OMa-1998, Subsection ISTC-3800 requires that instrumentation, including both measuring and test equipment and permanent plant instrumentation, used for valve testing and examination activities, shall (1) be properly controlled, calibrated, and adjusted in accordance with the owner's quality assurance program, and (2) have the accuracy, range, and repeatability characteristics necessary to verify compliance with the requirements of this subsection. In addition, instrumentation accuracy shall be considered when establishing valve test acceptance requirements.

NRC Recommendation

In OMb-2000 thru OM-2004 with 2005 and 2006 addenda, ISTC-3800, "Instrumentation," specifies requirements for instrumentation that is used in performing IST for check valve testing and examination activities. NRC staff expects that Instrumentation that is used in IST must be controlled and calibrated, and must have the accuracy, range, and repeatability necessary to verify compliance with the requirements. Accuracy and repeatability of the instrumentation are important considerations in the IST of safety-related check valves. IST should be performed in a way that permits the results to be compared for indications of valve degradation trends. When instrumentation that is used in valve testing and examination is not properly controlled and calibrated, any valve degradation indications may be masked, thereby diminishing the usefulness of the valve test results.

Basis for Recommendation

The instrumentation requirements and quality assurance activities specified in Subsection ISTC -3800 are needed to properly verify compliance with Code requirements and to detect and trend any operational degradation for assurance that the check valves will perform satisfactorily until the next IST. Appendix B to 10 CFR Part 50 provides for the quality assurance program and includes requirements for IST of safety-related components, test control, and control of measuring and test equipment.

4.1.10 Skid-Mounted Valves and Component Subassemblies

The exercising or examination of each check valve that is contained in (or part of) a skid-mounted package or major component is not always practical, particularly if the valve is enveloped within the structure of the package or component. Although the valve's performance may support a safety function, the practicality of exercising or examining each valve separately, as required by the OM Code, was not addressed prior to OMa-1996. Subsection ISTC-1200(c) of OMa-1996 exempts skid-mounted valves from the individual testing requirement and allows the valves to be tested as part of the overall package or major component. Specifically, Subsection ISTC-1200, states, in part, "Skid-mounted valves and component subassemblies are excluded from this Subsection, provided they are tested as part of the major component and are determined by the Owner to be adequately tested."

NRC Recommendation

Testing major components that include check valves that are an integral portion of the major component and that support the major component's performance of its safety function requires that the licensee must determine that the check valves are adequately tested. The NRC staff expects that, as part of the licensee's testing and determination responsibility, the check valves will be identified in the IST plan, along with an explanation of how the testing of the major component adequately tests the valves. The licensee should review the safety significance of the identified valves to ensure that the IST is adequate to demonstrate their continued operability. The documentation that provides assurance of the continued operability of the valves through the performed tests should be available at the plant site.

Basis for Recommendation

GDC 1, "Quality Standards and Records," in Appendix A to 10 CFR Part 50, in part, requires that components important to safety must be tested to quality standards commensurate with the importance of the safety functions performed.

4.1.11 Check Valves in New Reactors

RG 1.206, "Combined License Applications for Nuclear Power Plants (LWR Edition)," provides guidance for combined operating license (COL) licensees to develop IST programs for check valves. New nuclear power plants should provide the means for bi-directional testing of check valves within the scope of the IST program. As part of its review of design certification and COL applications, the NRC staff is evaluating whether new nuclear power plants will be designed to allow bi-directional testing of check valves.

Some new reactors include nozzle check valves in nuclear power plant systems to perform functions important to safety. Nozzle check valves are more complex than swing check valves used in most safety-related check valve applications in nuclear power plants. The provisions in the ASME OM Code focus on exercising check valves to demonstrate their operational readiness. The internal mechanism of the nozzle check valve might require more precise IST provisions to provide assurance of their operational readiness for their specific applications. Appendix A, "General Design Criteria for Nuclear Power Plants," to 10 CFR Part 50 states that where generally recognized codes and standards are used, they shall be identified and evaluated to determine their applicability, adequacy, and sufficiency, and shall be supplemented or modified as necessary to assure a quality product in keeping with the required safety function. Licensees of new reactors are responsible for satisfying 10 CFR Part 50, Appendix A, when developing IST provisions for nozzle check valves that provide reasonable assurance that they are capable of performing their safety functions.

The NRC staff will conduct inspections of the development and implementation of the IST program (including bi-directional testing of check valves and surveillance provisions for new check valve designs) during construction and operation of the new nuclear power plants.

4.2   Power-Operated Valves

Power-operated valves (POVs) are equipped with actuators that use motive force to change the position of the valve obturator. The types of actuators may include, for example, motor actuators, pneumatic actuators, hydraulic actuators, solenoid actuators, and pyrotechnic actuators. In addition, in the ASME OM Code, Subsection ISTC defines a power-operated relief valve (PORV) as a POV that can perform a pressure-relieving function and is remotely actuated by either a signal from a pressure-sensing device or controls switch and is not capacity certified under ASME B&PV Code, Section III, overpressure protection requirements. Certain valves, such as main steam isolation valves and valves that have a fail-safe function, may actuate open (or closed) on spring force. The ASME OM Code includes provisions for exercising, stroke-time testing, leak testing, and position-verification testing of POVs in the IST program. In the following sections, the NRC staff provides guidance concerning the implementation of specific Code provisions and associated regulatory requirements.

4.2.1   Stroke-Time Testing for Power-Operated Valves

In the ASME OM Code, Subsection ISTC-5113 specifies that active POVs shall have their stroke times measured when exercised in accordance with the nominal 3-month schedule specified in Subsection ISTC-3500. Stroke-time testing may indicate degradation in the performance of POVs. NRC requirements and guidance for supplementing the ASME Code provisions for stroke-time testing are discussed later in this document.

The ASME Code includes provisions for establishing reference values and limiting values for POV stroke times. Subsection ISTC-3300 states that reference values shall be determined from the results of preservice testing or inservice testing. Subsections ISTC-5113, 5121, 5131, 5141, and 5151 also state that limiting values for stroke time for various types of POVs shall be specified by the owner. In addition, Subsections ISTC-5114, 5122, 5132, 5142, and 5152 include a set of acceptance criteria for the reference value of the stroke time for POVs, and specify various corrective actions to be taken if those criteria are not satisfied. If the limiting value of stroke time is exceeded, Subsections ISTC-5115, 5123, 5133, 5143, and 5153 state that the POV shall be immediately declared inoperable.

The Code does not specify provisions for establishing the limiting value for stroke times, and it does not identify the relationship that should exist between those limits and the reference values for stroke time or any limits identified in the plant TS or safety analysis.

NRC Recommendation

The limiting value of full-stroke time should be based on the reference stroke time of a POV when it is known to be in good condition and operating properly. The limiting value should be a reasonable deviation from this reference stroke time, based on the size and type of the valve and power actuator. The deviation should not be so restrictive that it results in a POV being declared inoperable as a result of reasonable stroke time variations. However, the deviation used to establish the limiting value should be such that corrective action would be taken to provide assurance that the POV would remain capable of performing its safety function.

The limiting value for stroke time of a POV should be that point at which the licensee seriously questions continued operability. It is expected to be a value that is determined to be reasonable for the individual POV based on its characteristics and past performance, but not to exceed any safety analysis requirements. The value should not be based solely on the system requirements or values specified in safety analyses for system performance. When the identified limiting value is exceeded, the licensee shall declare the component inoperable and shall enter any applicable TS limiting condition for operation (LCO). After declaring the valve inoperable, the licensee should perform an analysis to identify the cause of the problem with the POV. If this analysis clearly demonstrates that the POV remains capable of performing its safety function, the analysis might constitute the corrective action required by the Code. The analysis must be documented.

Licensees should establish reference values that reflect the stroke time of the specific POV when in good condition and operating under applicable conditions. A licensee may establish additional sets of reference values as discussed in Subsection ISTC-3320, such as reference values that reflect test conditions of fluid pressure or flow in the system.

Licensees may use a quantitative multiplier on a reference time as a means of establishing a limiting value for stroke time. The licensee should document the justification for its selection of reference values for the stroke time of each POV, and should have this justification available at the plant site for review by NRC personnel.

Basis for Recommendation

The purpose of the limiting value of full-stroke time for a POV is to establish a value for taking corrective action on a degraded POV before it reaches the point where there is a high likelihood of failure to perform its safety function. While the TS provide assurance that important plant systems are capable of performing their safety functions in a timely manner during selected plant transient accidents and anticipated operational occurrences, the provisions of the ASME OM Code are intended to ensure the continued operability of particular plant components. The distinct bases for these two documents (i.e., TS and ASME Code) lead to criteria that may differ significantly. Nonetheless, the TS and ASME Code are both needed to provide confidence that the nuclear power plant can be operated safely. Therefore, licensees must follow the more restrictive criteria of the two documents, even though this might result in a component or system being declared inoperable. For example, if the TS or safety analysis limit for a POV is less than the IST value established using the above guidelines, the TS or safety analysis limit should be used as the limiting value of full-stroke time. When the TS or safety analysis limit for a POV is greater than the IST value established using the above guidelines, the limiting value of full-stroke time should be based on the above guidelines instead of the TS or safety analysis limit. The TS and safety analysis limits are useful for analyzing data when a POV has indicated degraded performance and been declared inoperable. In accordance with Subsections ISTC-5115, 5123, 5133, 5143, and 5153 the data may be analyzed to verify that the new stroke time represents acceptable POV operation.

4.2.2 Stroke-Time Measurements for Rapid-Acting Valves

In the ASME OM Code, Subsections ISTC-5114, 5122, 5132, 5142, and 5152 allow licensees to establish the limiting stroke time of 2 seconds for POVs that stroke in less than 2 seconds. The Code also eliminates the acceptance criterion related to the reference value for stroke-time for those POVs. However, new technologies and new applications of existing technologies enable licensees to time the strokes of rapid-acting valves with accuracy measured in milliseconds. Using new technology, licensees could establish an appropriate limiting stroke time based on a multiple of a reference value to ensure that corrective actions are taken if degrading conditions are identified.

NRC Recommendation

The NRC staff recommends that licensees should determine whether continued reliance on the 2-second limiting stroke time criterion in the ASME Code is appropriate when the actual stroke time can be measured in milliseconds.

Basis for Recommendation

The 2-second limiting stroke time for rapid-acting valves was based on measurement of stroke times using a stopwatch. Updated technology may improve the monitoring of the condition of these POVs or verify that a valve operates within a safety analysis limit that is less than 2 seconds.

4.2.3 Stroke Time for Solenoid-Operated Valves

The NRC is often asked to approve relief from the ASME Code provisions to allow licensees not to measure the stroke times of enclosed solenoid-operated valves (SOVs) that do not have position indication. If the licensee cannot time the stroke of an SOV by the conventional method using position indication, the licensee needs to propose a method to time the stroke of the valve or otherwise monitor the POV for degrading conditions to provide adequate assurance of its operational readiness. If the frequency provisions of the Code are met, the licensee does not need to request relief to use methods such as acoustics or diagnostic systems for stroke timing. If the licensee intends to apply a method to monitor for degradation other than by measuring stroke time, NRC authorization of the alternative is required pursuant to 10 CFR 50.55a(a)(3). For example, an enhanced maintenance program or periodic replacement may be acceptable when testing methods cannot be used effectively.

NRC Recommendation

The NRC staff recommends that licensees should use advanced diagnostic techniques to obtain stroke-time measurements in accordance with the frequency provisions of the Code, and should also use those advanced techniques or maintenance programs to monitor the degradation of SOV performance. In addition, the staff recommends that the technique should evaluate actual disk movement and not only movement of the pilot valve or valve stem.

Basis for Recommendation

In NUREG-1275, Vol. 6, "Operating Experience Feedback Report: Solenoid-Operated Valve Problems," the NRC described common-mode SOV problems that could significantly reduce plant safety. Several methods are available to measure stroke time or monitor the condition of SOVs using parameters such as the acoustic effects of disk movement, electric resistance, and the temperature of the coil. These advanced diagnostic techniques provide more precise means of monitoring SOV performance.

4.2.4 Supplement to the POV Stroke-Time Test Provisions of the ASME OM Code

Operational experience and valve testing programs have revealed weaknesses in the ability of stroke-time testing to assess the operational readiness of POVs to perform their safety functions. In response to those weaknesses, ASME, the NRC, and various industry groups have taken action to provide improved confidence in the capability of POVs to perform their safety functions under design-basis conditions.

With respect to motor-operated valves (MOVs), the NRC's regulations in 10 CFR 50.55a require that licensees who are implementing the ASME OM Code (beginning with the 1995 Edition) must supplement the provisions for MOV stroke-time testing specified in the Code with a program to ensure that the MOVs continue to be capable of performing their design-basis safety functions. In a *Federal Register* notice (64 FR 51370) dated September 22, 1999, the NRC discussed the implementation of MOV programs in response to GL 89-10, "Safety-Related Motor-Operated Valve Testing and Surveillance," (June 1989), and GL 96-05, "Periodic Verification of Design-Basis Capability of Safety-Related Motor-Operated Valves," (September 1996), as a means of satisfying the requirement to supplement MOV stroke-time testing.

The NRC established Generic Safety Issue (GSI) 158, "Performance of Safety-Related Power-Operated Valves Under Design-Basis Conditions," to evaluate whether additional regulatory actions were necessary to address performance issues for air-operated valves (AOVs), hydraulic-operated valve (HOVs), and solenoid-operated valves (SOVs). In Regulatory Issue Summary (RIS) 2000-03, "Resolution of Generic Safety Issue 158, 'Performance of Safety Related Power-Operated Valves Under Design-Basis Conditions'," dated March 15, 2000, the NRC closed GSI-158 on the basis that current regulations provide adequate requirements to ensure verification of the design-basis capability of POVs, and no new regulatory requirements are needed. In RIS 2000-03, the staff provided attributes for an effective POV testing program that incorporates lessons learned from MOV research and testing programs.

NRC Recommendation

The NRC staff recommends that licensees should apply their MOV programs established and implemented in response to GL 89-10 and GL 96-05 to supplement the provisions in the ASME OM Code for MOV stroke-time testing in accordance with the requirements in 10 CFR 50.55a. The staff also recommends that licensees consider information provided in RIS 2000-03, as well as lessons learned from their own MOV programs, to improve confidence in the capability of other POVs to perform their safety functions.

Basis for Recommendation

In the 1980s and 1990s, operating experience at nuclear power plants revealed that weaknesses in the ability of stroke-time testing to assess the operational readiness of POVs allowed performance deficiencies to remain undetected for an extended period of time. In 10 CFR 50.55a, the NRC requires that licensees whose code of record is the 1995 Edition (or a later edition or addenda) of the ASME OM Code must supplement their stroke-time testing of MOVs with programs to ensure that the MOVs are capable of performing their design-basis safety functions. As discussed in RIS 2000-03, the NRC staff determined that current requirements and guidance indicate the need for licensees to have confidence in the capability of all safety-related POVs to perform their design-basis functions. In RIS 2000-03, the staff discusses industry activities to improve POV performance.

In GL 89-10, the NRC asked licensees to ensure that MOVs in safety-related systems have the capability to perform their intended functions by reviewing MOV design bases, verifying MOV switch settings initially and periodically, testing MOVs under design-basis conditions where practical, improving evaluations of MOV failures and necessary corrective actions, and trending MOV problems. The NRC subsequently issued GL 96-05 to request that licensees establish a program, or ensure the effectiveness of their current program, to verify, on a periodic basis, that safety-related MOVs continue to have the capability to perform their safety functions within the current licensing basis of the facility. The NRC staff reviewed licensees' activities in response to GL 89-10 and GL 96-05 through plant-specific inspections and reviews of submitted information.

Licensees have completed their GL 89-10 programs for the operational nuclear power plants. In response to GL 96-05, the owners' groups of nuclear power plant licensees established the Joint Owners' Group (JOG) Program on MOV Periodic Verification as an industry-wide effort to evaluate potential degradation of MOV operating requirements. Most licensees committed to implement the JOG MOV program as part of their response to GL 96-05. In a safety evaluation dated September 2006 and its supplement dated September 2008, the NRC staff accepted the application of the JOG Program on MOV Periodic Verification as described in the safety evaluation and its supplement. As discussed in the safety evaluation, the JOG program addresses the operating requirements of valves and, therefore, licensees are responsible for justifying the output capability of MOVs within the scope of the JOG program. Further, licensees are responsible for justifying the long-term periodic verification programs for MOVs or their applications that are outside the scope of the JOG program. The staff considers the MOV design-basis capability verification performed by licensees in response to GL 89-10 and the MOV periodic verification program being conducted by licensees under GL 96-05, including any appropriate modifications in response to the JOG topical report and the NRC safety evaluation, to satisfy the requirement in 10 CFR 50.55a to supplement the MOV stroke-time test provisions of the ASME OM Code.

Although the NRC has not established new regulatory requirements to address performance issues for POVs other than MOVs, current NRC regulations and documents contain requirements and guidance intended to provide assurance that safety-related POVs are capable of performing their safety-related functions. For example, the regulations in Appendices A and B to 10 CFR Part 50 require that licensees must provide confidence that safety-related equipment (including POVs) is capable of performing its safety functions under design-basis conditions. Further, the regulations in 10 CFR 50.65, "Requirements for Monitoring the Effectiveness of Maintenance at Nuclear Power Plants," require that licensees must monitor the performance of structures, systems, and components (SSCs) in a manner sufficient to provide reasonable assurance that such SSCs are capable of fulfilling their intended functions. With respect to air systems, the NRC staff issued GL 88-14, "Instrument Air Supply System Problems Affecting Safety-Related Equipment," to ask licensees to verify (by test) that AOVs will perform as expected in accordance with all design-basis events. The staff provided the results of studies of POV issues in NUREG-1275, "Operating Experience Feedback Report," Volumes 2, 6, and 13; NUREG/CR-6644, "Generic Issue 158: Performance of Safety-Related Power-Operated Valves Under Operating Conditions," and NUREG/CR-6654, "A Study of Air-Operated Valves in U.S. Nuclear Power Plants." In RIS 2000-03, the staff provided a list of attributes of a successful POV design capability and long-term periodic verification program based on lessons learned from staff reviews of valve programs and plant visits. The staff also prepared several information notices to alert licensees to IST issues related to POV performance. Specifically, these included IN 86-50, "Inadequate Testing To Detect Failures of Safety-Related Pneumatic Components or Systems; IN 85-84, "Inadequate Inservice Testing of Main Steam Isolation Valves;" and IN 96-48, "Motor-Operated Valve Performance Issues," which described lessons learned from MOV programs that are applicable to other POVs.

ASME has initiated efforts to improve the Code provisions for assessing the operational readiness of POVs in IST programs at nuclear power plants. For example, ASME has incorporated ASME OM Code Cases OMN-1, "Alternative Rules for Preservice and Inservice Testing of Certain Electric Motor-Operated Valve Assemblies in LWR Power Plants," and OMN-11, "Risk Informed Testing of Motor-Operated Valves," as Appendix III in the 2009 Edition to the ASME OM to replace quarterly MOV stroke-time testing with periodic exercising and diagnostic testing. (Note: The details related to the ASME OM-2009 are for information only.) ASME also developed OM Code Case OMN-12, "Alternate Requirements for Inservice Testing Using Risk Insights for Pneumatically and Hydraulically Operated Valve Assemblies in Light-Water Reactor Power Plants (OM Code 1998, Subsection ISTC)," that provides guidance for an alternative to quarterly stroke-time testing for AOVs and HOVs. OMN-12 includes risk-informed provisions to allow licensees to obtain more precise performance data for use in assessing the operational readiness of POVs that are determined to have higher safety significance, while allowing licensees to obtain less precise data for POVs of lower safety significance.

In addition to the NRC and ASME, the nuclear industry is taking action to address POV performance issues. As discussed above, the industry developed the JOG Program on MOV Periodic Verification to share resources among licensees and to establish an improved response to the MOV issues to be addressed under GL 96-05. In addition, a Joint Owners' Group on Air-Operated Valves (JOG AOV) has established a voluntary program to improve confidence in the capability of safety-related AOVs to perform their design-basis functions. In RIS 2000-03, the NRC staff noted that it provided comments on the JOG AOV program in a letter to the NEI dated October 8, 1999. The staff also stated that it would continue to monitor licensees' activities to ensure that POVs are capable of performing their safety-related functions under design-basis conditions. If the industry does not adequately address POV functionality under design-basis conditions, the NRC staff indicated in RIS 2000-03 that additional regulatory action may be necessary.

4.2.5 Alternatives to POV Stroke-Time Testing

Given the weakness in stroke-time testing for assessing the operational readiness of POVs, ASME developed alternatives to the Code provisions for stroke-time testing of POVs. As an alternative to MOV stroke-time testing, ASME developed Code Case OMN-1, which provides periodic exercising and diagnostic testing for use in assessing the operational readiness of MOVs. In Code Case OMN-11, ASME provides additional guidance for use with Code Case OMN-1 to emphasize the testing provisions for MOVs in the IST program that are determined to have high safety significance, while allowing less precise testing for MOVs that are determined to have lower safety significance. ASME has incorporated these code cases into Appendix III to the 2009 Edition of the ASME OM Code to replace quarterly MOV stroke-time testing with periodic exercising and diagnostic testing. (Note: The details related to the ASME OM-2009 are for information only.)

With respect to AOVs and HOVs, ASME developed Code Case OMN-12 to provide an alternative to the Code stroke-time testing provisions that incorporates risk insights to focus on AOVs and HOVs in the IST program that are determined to have the highest safety significance, while allowing less emphasis on AOVs and HOVs that have lower safety significance.

NRC Recommendation

The NRC regulations in 10 CFR 50.55a(b)(3) require nuclear power plant licensees implementing the ASME OM Code incorporated by reference in 10 CFR 50.55a to supplement the quarterly MOV stroke-time testing provisions in the ASME OM Code with a program to periodically demonstrate the design-basis capability of safety-related MOVs. The NRC staff considers that the MOV programs developed in response to GLs 89-10 and 96-05 together will satisfy the regulatory requirement to supplement the MOV stroke-time provisions in the ASME OM Code. The staff also considers the provisions in ASME OM Code Case OMN-1 as accepted in RG 1.192 to satisfy the regulatory requirement to supplement the quarterly MOV stroke-time provisions. ASME has replaced the quarterly MOV stroke-time testing provisions in Appendix III to the 2009 Edition of the ASME OM Code that incorporates the provisions in ASME OM Code Cases OMN-1 and OMN-11. (Note: The details related to the ASME OM-2009 are for information only.) The NRC staff also considers the alternative approach to quarterly stroke-time testing of AOVs and HOVs in ASME OM Code Case OMN-12 as accepted in RG 1.192 to provide an acceptable alternative to the ASME OM Code stroke-time testing provisions for those POVs. Therefore, the NRC staff recommends that licensees implement

ASME Code Cases OMN-1, OMN-11, and OMN-12, as accepted by the NRC (with certain conditions) in RG 1.192, as alternatives to the stroke-time testing provisions in the ASME Code for applicable POVs. When the 2009 Edition to the ASME OM Code is incorporated by reference in the NRC regulations, licensees will be responsible for modifying their MOV testing programs in accordance with the 10-year update requirements for IST programs. (Note: The details related to the ASME OM-2009 are for information only.)

Basis for Recommendation

As part of the ASME Code Committee process, industry experts have developed ASME Code Cases to address weaknesses in the ability of stroke-time testing to assess the operational readiness of POVs in IST programs at nuclear power plants. The ASME Code Cases incorporate risk insights to emphasize IST provisions for POVs that are determined to have the highest safety significance. The NRC has reviewed and accepted several of these Code Cases with certain conditions. For example, RG 1.192 allows licensees with an applicable code of record to implement ASME Code Case OMN-1 (in accordance with the provisions in the regulatory guide) as an alternative to the Code provisions for MOV stroke-time testing, without submitting a request for relief from their code of record. In RG 1.192, the staff also accepts (with certain conditions) the use of the risk-informed provisions in Code Case OMN-11 by applicable licensees, in conjunction with Code Case OMN-1. RG 1.192, "Operation and Maintenance Code Case acceptability, ASME OM Code," also allows licensees with an applicable code of record to implement Code Case OMN-12 for AOVs and HOVs (with certain conditions) in lieu of the Code provisions for stroke-time testing, without the need to submit a relief request. Licensees with a code of record that is not applicable to the acceptance of these Code Cases may submit a request to apply those Code Cases, consistent with the indicated conditions, as an alternative to the ASME OM Code that provides an acceptable level of quality and safety. ASME has incorporated ASME OM Code Cases OMN-1 and 11 as Appendix III to the 2009 Edition of the ASME OM Code. Therefore, the NRC regulations will require licensees to modify their MOV testing programs to apply the provisions of Appendix III to the ASME OM Code in accordance with the 10-year update requirements for IST programs. (Note: The details related to the ASME OM-2009 are for information only.)

4.2.6 Main Steam Isolation Valves

In IN 85-84, "Inadequate Inservice Testing of Main Steam Isolation Valves (MSIVs)," the NRC staff described an inadequacy in the IST of MSIVs. Specifically, the staff stated that two different licensees were testing their MSIVs using the nonsafety-related instrument air to achieve closure. Fail-safe IST of MSIVs as required by Subsection ISTC-3560 necessitates the removal of the instrument air supply and electric power. Concerns related to MSIVs are described in IN 94-08, "Potential for Surveillance Testing To Fail To Detect an Inoperable Main Steam Isolation Valve," and IN 94-44, "Main Steam Isolation Valve Failure To Close on Demand Because of Inadequate Maintenance and Testing."

NRC Recommendation

The staff recommends that licensees review their inservice and fail-safe testing to ensure compliance with Code requirements.

Basis for Recommendation

The practice of performing IST of components that are relied on to mitigate the consequences of accidents using sources of power that were not considered in the safety analyses is inconsistent with the objective of periodic IST for fail-safe testing. In IN 85-84, the NRC staff alerted licensees that, with low or no steam flow, the MSIV might not close unless instrument air is available to power the actuator.

In its Service Information Letter 477, the General Electric Company (GE) described a related concern for BWRs in which excessive tightening of gland flanges in the MSIV can prevent the valve from closing in response to spring force alone. During a postulated design-basis accident in which a recirculation line breaks with the MSIVs open, containment pressure may increase significantly, and may exert an opening force on the valve actuators inside containment. Under such circumstances, the MSIV springs alone will not close the MSIV unless the spring force can overcome the combination of the opening force caused by containment pressure and the resistive force caused by stem packing friction. GE recommended a review of packing chamber maintenance practices, "springs-only" full-stroke closing tests, a force balance in which containment pressure is considered, a leak-tightness test of the MSIV actuator and accumulator, and a modification of the applicable licensing-basis documents. GE noted that this would necessitate measurement of the actual valve stem travel because the final 10 percent of stem travel coincides with the weakest spring force. GE stated that, by monitoring position switches alone, a utility could not determine that the valve is fully closed because the switches monitor the valve only when it is 90-percent open or 90-percent closed. One BWR licensee identified that the MSIVs would not pass local leak rate testing after closing on spring force only.

4.2.7 Verification of Remote Position Indication for Valves by Methods Other Than Direct Observation

ASME OM Code (2004 Edition), Subsection ISTC-3700, "Position Verification Testing," states

> Valves with remote position indicator shall be observed locally at least once every 2 years to verify that valve operation is accurately indicated. Where practicable, this local observation should be supplemented by other indications such as the use of flowmeters or other suitable instrumentation to verify obturator position. These observations need not be concurrent. Where local observation is not possible, other indications shall be used for verification of valve operation.

The requirement that valves with remote position indicators must be observed at least once every 2 years to verify that valve position is accurately indicated has been specified in the ASME OM Code and the previous ASME/ANSI OM Part 10 for many years. The additional guidance for supplementing the local observation was not present in older versions of OM Part 10, such as in the 1983 versions.

Many valves have no provision for verifying the position by direct observation. To verify the position by observation, licensees can disassemble the valve, which could introduce additional valve failure mechanisms. Other methods (such as nonintrusive techniques, causing the flow to begin or cease, leak testing, and pressure testing) can yield a positive indication of position.

NRC Recommendation

ISTC-3530, "Valve Obturator Movement," allows obturator movement to be determined by indicating lights in the control room when exercising the valve to meet the quarterly stroke-time testing requirement of the ASME OM Code. The valve position verification testing required by ISTC-3700 provides confirmation on a 2-year frequency that the indicating lights reflect actual valve operation. ISTC-3700 allows flexibility to licensees in verifying that operation of valves with remote position indicators is accurately indicated. Operating experience has revealed that indicating light might not be sufficient to verify valve position. The extent of verification necessary for valve operation to satisfy ISTC-3700 will depend on the type of valve, the sophistication of the diagnostic equipment used in testing the valve, possible failure modes of the valve, and the operating history of the valve and similar valve types. After such consideration, the licensee will be responsible for developing and implementing a method to verify that valve operation is accurately indicated as required by ISTC-3700.

Basis for Recommendation

Appendix A to 10 CFR Part 50 states that where generally recognized codes and standards are used, they shall be identified and evaluated to determine their applicability, adequacy, and sufficiency, and shall be supplemented or modified as necessary to assure a quality product in keeping with the required safety function. Nuclear power plant licensees are responsible for satisfying 10 CFR Part 50, Appendix A, where codes and standards are insufficient to provide reasonable assurance that components are capable of performing their safety functions.

In Section 4.2.5, "Verification of Remote Position Indication for Valves by Methods Other Than Direct Observation," of NUREG-1482, "Guidance for Inservice Testing at Nuclear Power Plants," dated April 1995, the NRC staff noted that the Code requires that valves with remote position indicators be observed at least once every 2 years to verify that valve position is accurately indicated. The staff stated that if remote valve position cannot be verified by local observation at the valve, an acceptable approach is for the licensee to observe operational parameters such as leakage, pressure, and flow that give positive indication of the valve's actual position. The staff indicated its interpretation of the Code requirement by stating that for certain types of valves that can be observed locally, but for which valve stem travel does not assure the stem is attached to the disk, the local observation must be supplemented by observing an operating parameter as required in the Code. In the basis discussion, the staff stated that accurate position indication for safety-related valves is important for reactor operation during all plant conditions. Therefore, the staff noted that the Code requires verification of the accuracy of the remote position indication for all valves in the IST program with remote position indication. The staff indicated that many positive ways are available to verify the indication that a valve is open or closed. For example, the staff referenced leak-rate testing, in-line flow rate instrumentation, and system and differential pressures for indication of valve position.

In Section 4.2.7 of Revision 1 to NUREG-1482, dated January 2005, the NRC staff discussed its interpretation of the ASME OM Code for position indication verification. The staff noted that ISTC-3700 requires verification of the accuracy of the remote position indication for all valves in the IST program with remote position indication. The staff indicated that if licensees cannot verify remote valve position by local observation at the valve, an acceptable approach is for the licensee to observe operational parameters (such as leakage, pressure, and flow) that give a positive indication of the valve's actual position. In Revision 1, the staff modified the discussion on supplementing local observation of valves for which stem travel does not assure that the stem is attached to the disk with observation of operating parameters (i.e., changing "must" to "should") because Revision 1 to the NUREG removed the preapproval of this approach and required Commission approval pursuant to 10 CFR 50.55a(f)(4)(iv).

The Code requires licensees to verify the accuracy of the remote position indication for all valves in the IST program that have remote position indication. Subsection ISTC-3700 states that where local observation is not possible, licensees shall use other indications to verify operation. Nuclear power plant operating experience has revealed that reliance on indicating lights and stem travel are not sufficient to satisfy the requirement in ISTC-3700 to verify that valve operation is accurately indicated for those valves where the integrity of the internal mechanism of the valve (such as stem-to-disk connection) cannot be assured. Criteria V of Appendix B to 10 CFR Part 50 requires safety-related components that are subjected to test activities be required to have appropriate instructions, procedures, or drawings and qualitative or quantitative acceptance criteria for determining that activities has been successfully completed. Therefore, licensees are responsible for developing a method to verify that valve operation is accurately indicated to satisfy ISTC-3700 requirements such that IST can help to identify the stem-to-disc separation as valves are tested.

The NRC staff discussed the ASME OM Code provisions for valve position verification in IN 2012-14 (July 24, 2012), "Motor-Operated Valve Inoperable due to Stem-Disc Separation."

4.2.8   Requirements for Verifying Position Indication of Passive Valves

The Code does not restrict the verification of position indication to only active valves. The ASME OM Table ISTC 3500-1 indicates that the licensee must also locally verify the position indication for Category B passive valves. As discussed in Section 4.2.7 of this NUREG, the licensee is responsible for developing and implementing a method to verify that valve operation is accurately indicated as required by ISTC-3700. The extent of verification necessary for valve operation to satisfy ISTC-3700 will depend on the type of valve, the sophistication of the diagnostic equipment used in testing the valve, possible failure modes of the valve, and the operating history of the valve and similar valve types.

The Code does not require licensees to verify the indication at the remote panels. However, verification at remote panels is a good practice and provides confidence in the remote indication.

4.2.9   Control Valves with a Safety Function

In general, control valves that used only for system control would be exempt from IST as discussed in Subsection ISTC-1200. However, some control valves also perform safety or fail-safe functions (e.g., fail open, fail closed, fail as-is), and such valves must be tested in accordance with the requirements for IST. The staff has received many requests for relief from stroke-time measurement requirements, based on the impracticality of performing the measurement by the conventional method using position indication lights. Typically, the control valves do not have position indication, and testing can only be performed by bypassing control signals. To allow stroke timing by bypassing the control signals of those control valves that have position indication lights, the licensee may have to drain systems, which might make it impractical to test at the Code-defined frequency.

NRC Recommendation

Control valves that perform a safety or fail-safe function must be tested in accordance with the Code provisions for IST to monitor the valves for degrading conditions. The NRC staff recommends that licensees should apply ASME Code Case OMN-8, as accepted in RG 1.192, if concerns exist regarding IST of control valves with fail-safe functions. Code Case OMN-8 states that stroke-time testing need not be performed for POVs when the only safety-related function of those valves is to fail safe. Any abnormality or erratic action experienced during valve exercising should be recorded in the test record and an evaluation should be performed.

4.2.10 Pressurizer Power-Operated Relief Valve Inservice Testing

Power-operated relief valves (PORVs) were often not purchased as safety-related valves, and the function of these valves to provide pressure control for plant transients was not considered safety-related. The valves were not designed to serve as overpressure protection devices during power operations, as required by ASME B&PV Code, Section III, but many have been used as low-temperature overpressure protection valves.

NRC Recommendation

Recognizing that the PORVs have shown a high likelihood of sticking open and are not needed for overpressure protection during power operation, the provisions in Subsections ISTC-3500 and ISTC-5100 for exercising quarterly during power operation are not practical and, therefore, exercising may be performed during cold shutdown conditions. Subsection ISTC-3310 requires licensees to perform testing after maintenance or repair. Test methods must confirm that the PORV has been reassembled correctly and is capable of performing its design function. There have been instances (see IN 96-02, "Inoperability of PORVs Masked by Downstream Indications During Testing") where improper evaluation of testing failed to identify the incorrect reassembly of a PORV.

Previously approved NRC guidance included in GL 90-06 (see below) indicates that, because the PORVs function during reactor startup and shutdown to protect the reactor vessel and coolant system from low-temperature overpressurization conditions, they should be exercised before system conditions warrant vessel protection, and should also be exercised after the operational readiness of the block valves is ensured, by exercising and stroke-timing according to the following test schedule:

- Perform full-stroke exercising during each cold shutdown or, as a minimum, once each refueling cycle.
- Perform stroke timing during each cold shutdown, or as a minimum, once each refueling cycle.
- Perform fail-safe testing during each cold shutdown, or as a minimum, once each refueling cycle.

- Include the PORV block valves in the IST program, and test them quarterly to ensure protection against a small-break LOCA in the event that a PORV fails open.
- If the plant frequently enters cold shutdown mode, testing of the PORVs is not required more often than once every 3 months,

Basis for Recommendation

The NRC's guidance on the IST requirements for PORVs is included in GL 90-06, "Resolution of Generic Issue 70: Power-Operated Relief Valve and Block Valve Reliability, and Generic Safety Issue 94: Additional Low-Temperature Overpressure Protection for Light-Water Reactors, pursuant to 10 CFR 50.54(f)." In IN 89-32, "Surveillance Testing of Low-Temperature Overpressure-Protection Systems," the NRC discussed the stroke time assumptions made in plants' safety analyses for these PORVs, and the IST performed for these valves. Stroke times of the valves were unacceptable or were not measured in the direction required for low-temperature overpressure systems to prevent exceeding the limits in Appendix G to 10 CFR Part 50. Compliance with the guidance in GL 90-06 has been coordinated between the plants and the NRC project managers for each plant on a case-by-case basis.

4.2.11 Online Check Valve Sample Disassembly and Inspection

Licensees have proposed, as an alternative to ISTC-5221(c) and ISTC-5224, to perform sample disassembly and inspection of check valves in a group online. Subsection ISTC of OM Code, Paragraph ISTC-3510, requires that check valves be exercised every 3 months. Paragraph ISTC-3522(c) states that if exercising is not practicable during operation at power and cold shutdown, it shall be performed during refueling outages. ISTC-5221(c) allows disassembly of check valves every refueling outage as an alternative means to verify their operability. Instead of disassembly every refueling outage, ISTC-5221(c) provides the option of using a sample disassembly and inspection program for groups of identical valves in similar applications. Further, ISTC-5221(c)(3) states that at least one valve from each group shall be disassembled and examined at each refueling outage and all valves in each group shall be disassembled and examined at least once every 8 years. ISTC-5224 requires that check valves in a sample disassembly program that are not capable of being full-stroke exercised or have failed or have unacceptably degraded valve internals, shall have the cause of failure analyzed and the condition corrected.

ISTC-5224 also states that other check valves in the sample group that may also be affected by this failure mechanism be examined or tested during the same refueling outage to determine the condition of internal components and their ability to function. A licensee should fully describe how it plans to comply with the requirements in ISTC-5224 when submitting alternative requests for check valve group sample disassembly and inspection online. The plan description also should include information on management of examination and testing of all group valves should a scheduled valve inspection be declared inoperable. For example, licensees should explain how the disassembly and inspection of the other check valves in a group will be completed within the allowed system outage time.

4.2.12 POVs in New Reactors

RG 1.206 provides guidance for COL licensees to develop IST programs for POVs including, for example, MOVs, air-operated valves, hydraulic-operated valves, and solenoid-operated valves. The NRC staff reviews the consideration of diagnostic testing of POVs as part of the review of IST program descriptions provided in support of COL applications. The NRC regulations in 10 CFR 50.55a require licensees of new nuclear power plants to update their IST programs to the most recent edition and addenda of the ASME OM Code incorporated by reference in the NRC regulations 12 months before fuel loading. The staff will conduct inspections of the development and implementation of the IST program (including POV diagnostic testing) during construction and operation of the new nuclear power plants.

With respect to MOVs, the NRC regulations in 10 CFR 50.55a(b)(3)(ii) will require licensees of new reactors to supplement the quarterly MOV stroke time testing provisions in the ASME OM Code with a program to periodically demonstrate the design-basis capability of safety-related MOVs. COL licensees may apply the JOG Program on MOV Periodic Verification as accepted in an NRC safety evaluation dated September 2006 and its supplement dated September 2008 as part of satisfying 10 CFR 50.55a(b)(3)(ii). As discussed in the safety evaluation, the JOG program addresses the operating requirements of valves and, therefore, licensees are responsible for justifying the output capability of MOVs within the scope of the JOG program. Further, licensees are responsible for justifying the long-term periodic verification programs for MOVs or their applications that are outside the scope of the JOG program. If the 2009 Edition (or later edition or addenda) of the ASME OM Code has been incorporated by reference in the regulations, the new reactor licensee will be required to implement Appendix III to the ASME OM Code that replaces quarterly MOV stroke-time testing with periodic exercising and diagnostic testing. (Note: The details related to the ASME OM-2009 are for information only).

With respect to POVs other than MOVs, the NRC staff evaluates whether the IST program description includes the ASME OM Code provisions for quarterly stroke-time testing. The staff also evaluates whether IST program description for POVs other than MOVs incorporates the lessons learned from MOV testing and research programs. For example, RIS 2000-03 includes attributes of a successful IST program for POVs that incorporates lessons learned from MOV operating and testing experience. ASME is developing improved IST provisions for POVs other than MOVs for incorporation into the ASME OM Code. See Section 4.4.8 for IST guidance regarding pyrotechnic actuated valves.

4.3   Safety and Relief Valves

4.3.1   Scope

Subsection ISTA-1100 defines the scope of the valves subject to IST to include pressure-relief devices that protect systems (or portions of systems) that perform a required function in shutting down the reactor to the safe shutdown condition, maintaining the safe shutdown condition, or mitigating the consequences of an accident that results from overpressure. Pressure-relief valves, which are installed in systems to protect against overpressure, may not, of themselves, appear to perform a specific function to shut down the reactor, maintain it in a safe shutdown condition, or mitigate the consequences of an accident. (Automatic depressurization valves in BWRs are an example of relief valves that perform both an overpressure protection function and a function to depressurize the primary system when opened on an automatic signal or by an operator.) However, they may be required to be included in the IST program and tested according to the schedules stipulated in Subsection ISTC and Appendix I of the ASME OM Code. Specifically, Subsection ISTC of the ASME OM Code clarifies that its requirements apply only to pressure-relief devices required for overpressure protection. The testing of these devices is to be included in 120-month updated IST programs.

Testing of "thermal relief valves" has been the subject of much discussion over the past several years. Contributing to some confusion, in many original system designs, so-called thermal relief valves were installed to protect isolated segments of piping that could be pressurized as a result of heating from some source, but were widely viewed as having no safety-related function in mitigating the consequences of accidents or ensuring any other system safety function. GL 96-06, "Assurance of Equipment Operability and Containment Integrity During Design-Basis Accident Conditions," emphasized the importance of protecting certain isolated segments of piping from excessive thermally-induced pressurization, especially where containment integrity could be affected. In recent years, ASME has made changes to Appendix I to the OM Code to include specific requirements to periodically test or replace thermal relief valves.

The requirement to test safety and relief valves (S/RVs) that provide overpressure protection is based on the requirements of Section III of the ASME B&PV Code, as well as the *USA Standard Code for Pressure Piping* (USAS B31.1) and the *USA Standard Code for Nuclear Power Piping* (USAS B31.7). If the results of an overpressure protection "re-analysis" for a particular system indicate that a relief valve is not necessary, it may be removed from the scope of the IST program.

As required by ASME B&PV Code, Section III, Article NX-7200, it is the Owner's responsibility to prepare, certify and file an Overpressure Protection Report for the facility. The Overpressure Protection Report defines the protected systems and the integrated overpressure protection provided. Article NX-7200 also contains requirements regarding verification that pressure relief devices are not required and reconciliation of the Overpressure Report following modifications.

4.3.2   Method of Testing Safety and Relief Valves

4.3.2.1   BWR Safety/Relief Valve Stroke Testing

In recent years, the NRC staff has received numerous requests for relief or TS changes or both related to the stroke testing requirements for BWR dual-function main steam S/RVs. The 2003 Addendum and earlier editions and addenda to Mandatory Appendix I to the OM Code require the stroke testing of S/RVs after they are reinstalled following maintenance activities. A number of licensees have determined that in situ testing of the S/RVs can contribute to undesirable seat leakage of the valves during subsequent plant operation and have received approval to perform stroke testing at a laboratory facility coupled with in situ tests and other verifications of actuation systems as an alternative to the testing required by the OM Code. The revised subparagraph I-3410(d) in Mandatory Appendix I to the 2004 Edition of the OM Code does not require licensees to stroke test S/RVs at reduced or normal system pressure following maintenance. Subparagraph I-3410(d) in the 2004 Edition of the OM Code requires that each S/RV that has been removed for maintenance or testing and reinstalled shall have the electrical and pneumatic connections verified either through mechanical/electrical inspection or testing before the resumption of electric power generation. Several licensees have requested and obtained NRC approval in accordance with 10 CFR 50.55a(f)(4)(iv) to use Subparagraph I-3410(d) of the 2004 Edition of the OM Code in place of Subparagraph I-3410(d) of the 2001 Edition through the 2003 Addenda to the OM Code.

4.3.2.2   PWR Main Steam Safety Valve Set Pressure Testing

To reduce the need to remove valves from their installed position and the time required to perform set pressure testing, many PWR licensees perform testing of main steam safety valves (MSSVs) using in situ testing equipment with operating steam pressure. One advantage of this method is that actual environmental and fluid temperature conditions are used, in lieu of duplicating them in a test laboratory. However, this method has introduced inaccuracies because the set pressure is determined by a combination of the measured system operating pressure and the applied assisting force provided by the testing device. This assisting force is applied by pneumatic pressure on a piston or diaphragm and is converted to an equivalent additional amount of system pressure by dividing the force by the valve disk area against which the system pressure acts. Inaccuracies in the value of the disk area have caused some inaccuracies in the set pressure determination, as discussed in IN 94-56, "Inaccuracy of Safety Valve Set Pressure Determination Using Assist Devices."

4.3.3   Jack-and-Lap Process

In IN 91-74, "Changes in Pressurizer Safety Valve Setpoints Before Installation," the NRC stated that the setpoint changes in Dresser pressurizer safety valves could result, in part, from changes made during a jack-and-lap procedure that is performed after setpoint testing and before installation to reduce seat leakage. This procedure may have lacked adequate controls.

Many licensees avoid performing setpoint testing after jack-and-lap maintenance because this testing could lead to leakage. Subsection ISTC and Appendix I of the OM Code require that after repairing a valve or performing maintenance that could affect the valve's performance, the licensee must demonstrate that the performance parameters are acceptable by testing the valve before returning it to service. The licensee must test pressure relief devices as required by Subsection ISTC and Appendix I following replacement, repair, and maintenance.

The staff recommends that, if a licensee chooses to use the jack-and-lap process and not re-verify the set pressure of the valves, the licensee must determine whether the maintenance activity is of an extent that a setpoint test is required after the valve is reassembled and reinstalled. If the jack-and-lap process is controlled so that the setpoint will not be affected, the licensee may not need to perform a test. Action in accordance with this recommendation necessitates determination of the effect of this activity and evaluation within the quality controls and quality assurance for the process. Controls include limits on the amount of material that is removed, the controls to ensure that the settings and adjustments of the valve parts that affect the setpoint are not changed, and the requirements in the maintenance procedure to address any unusual conditions that occur during the maintenance activity. The licensee may also consider industry experience to determine whether changes in the methods of performing this activity are necessary as plants accumulate more data. Because the NRC staff cannot make this determination by evaluating a relief request, relief is neither appropriate nor available for this activity.

4.3.4 Maintenance and Inspection of Safety and Relief Valves in Addition to OM Code Requirements

Licensees should note that not all maintenance and inspection that may be needed to ensure continued functional capability of safety, relief, and pilot valves is necessarily performed as a result of inservice testing required by the OM Code. In a recent case involving some BWR S/RVs, additional periodic maintenance and inspection of certain internal parts were necessary to check for excessive wear and eventual binding of the main disks. This was discovered on valves that had successfully passed required inservice tests and is discussed further in IN 2003-01, "Failure of a Boiling-Water Reactor Main Steam Safety/Relief Valve."

4.3.5 Scheduling of Safety and Relief Valve Testing

Appendix I to the OM Code requires that licensees must test a minimum size sample of valves within a valve group within a specified period. A penalty is also applied, in that additional valves must be tested when any of the samples fail to meet the necessary acceptance criteria. In determining the minimum acceptable sample size, fractions of valve numbers resulting from calculating the number of valves to be tested are to be rounded to the next higher whole number.

4.3.6 Use of ASME OM Code Case OMN-17

Many licensees have requested and obtained NRC authorization in accordance with 10 CFR 50.55a(a)(3) to use the provisions in Code Case OMN-17, "Alternative Rules for Testing ASME Class 1 Pressure Relief/Safety Valves," as an alternative to the 5-year test interval specified in the OM Code. Code Case OMN-17 allows an extension of the test frequency for S/RVs from 60 months to 72 months plus a 6-month grace period. The code case imposes a special maintenance requirement to disassemble and inspect each valve to verify that parts are free from defects resulting from the time-related degradation or maintenance-induced wear before the start of the extended test frequency. Although the OM Code does not require that S/RVs be routinely refurbished, refurbishment provides reasonable assurance that the S/RVs are operationally ready during the extended test interval. ASME published Code Case OMN-17 in the 2009 Edition of the OM Code. The NRC will consider including Code Case OMN-17 in a future revision to RG 1.192. Code Case OMN-17 will not be acceptable for use without prior NRC review and approval unless the Code Case is included in RG 1.192. Until such inclusion, licensees must submit an alternative request to the NRC and obtain authorization to use Code Case OMN-17.

4.4 Miscellaneous Valves

The following issues and NRC recommendations apply to miscellaneous types of valves.

4.4.1 Post-Accident Sampling System Valves

NUREG-0737, Clarification of TMI Action Plan Requirements, Section II.B.3, details the requirements and capabilities of post-accident sampling systems (PASSs) for sampling both the reactor coolant and the containment atmosphere. The PASS consists of pumps and valves that perform these and possibly other functions. The PASS also includes valves that perform a containment isolation function.

NRC Recommendation

The IST program applies to any PASS valves within the scope of 10 CFR 50.55a and the ASME OM Code. Such valves in the PASS that perform a containment isolation function must be included in the IST program as Category A or A/C and must be tested to Code requirements except where relief has been granted.

The remaining valves in the PASS would typically be tested as required by the TS or other documents and need not be included in the IST program. However, the staff recommends that if the licensee elects to include these valves in the IST program, a note should be included that the testing is beyond the scope of 10 CFR 50.55a.

In many cases, a licensee's TS have been amended to eliminate the requirements to have and maintain a PASS. If a PASS valve is eliminated from the TS but still performs a function within the scope of 10 CFR 50.55a and the ASME OM Code, the valve should remain in the IST program.

4.4.2 Post-Maintenance Testing After Stem Packing Adjustments and Backseating of Valves to Prevent Packing Leakage

Subsection ISTC-3310, "Effects of Valve Repair, Replacement, or Maintenance on reference Values," requires that, when a valve or its control system has been replaced, repaired, or has undergone maintenance that could affect the valve's performance, a new reference value shall be determined or the previous values reconfirmed by an inservice test before the time it is returned to service or immediately if not removed from service.

Examples of maintenance are provided in a footnote of ISTC-3310 and include: adjustment of stem packing, limit switches, or control system valves, removal of the bonnet, stem assembly, actuator, obturator, or control system components.

Backseating a valve may also affect its performance (e.g., cause damage to the valve or bind it into its back seat).

It may be necessary to adjust the stem packing during power operations in order to stop stem packing leaks on valves that must remain in position for operations to continue. Examples include MSIVs and main feedwater isolation valves. If the leakage does not pose a personnel safety hazard, licensees may adjust the packing without removing the valves from service. Alternatively, backseating a valve may stop packing leakage without the need to take the valve out of service. Licensees should exercise caution when performing such maintenance, as improper backseating or adjustment of valve stem packing could adversely affect the valve's functional capability.

NRC Recommendation

The staff has determined that it is acceptable for licensees to perform an engineering evaluation of the impact of adjusting valve stem packing or backseating a valve to demonstrate that the performance parameters are within acceptable limits if a stroke test cannot be performed under current plant conditions. If it is necessary to adjust the stem packing or backseat a valve to stop packing leakage and if a required stroke test or leak rate test is not practical in the current plant mode, the licensee must, at a minimum, justify by analysis that (1) the packing adjustment is within manufacturer-specified torque limits for the existing packing configuration, (2) the backseating does not deform the valve stem, and (3) the performance parameters of the valve are not adversely affected. In addition, the licensee must perform a confirmatory test at the first available opportunity when plant conditions allow testing. Packing adjustments beyond the manufacturer's limits may not be performed without (1) an engineering analysis showing that the performance parameters of the valve are not adversely affected, and (2) input from the manufacturer, unless tests can be performed after adjustments.

Examples of such valves are MSIVs and main feedwater isolation valves, which must remain open to continue power operations. The licensee must evaluate any data available from previous testing with the packing torqued to the specified limit, and must verify that the valve was leak tight and previously stroked within acceptable limits with the packing adjusted to the higher value, or from previous instances of backseating a valve.

NRC grant of relief under 10 CFR 50.55a(f) is not necessary because this action is in accordance with the Code requirements if the licensee can demonstrate that the performance parameters will not be adversely affected.

To properly implement this guidance, licensees must perform a partial-stroke test, if practical, to obtain further assurance that the valve stem is free to move. At the first opportunity when the plant enters an operating mode in which testing is practical, the licensee must test all valves that have had packing adjustments or been backseated without post-maintenance testing. The maintenance procedure used to adjust the packing must include the limits, and any changes to the torque limits are subject to a 10 CFR 50.59, "Changes, Tests, and Experiments," review. Licensees should avoid adjusting redundant valves without performing post-maintenance testing. Backseating procedures should include precautions to prevent stem deformation.

To properly implement this guidance, a licensee must evaluate valves individually, unless it has established a valve packing program in which designated limits, justified by test data, allow adjustments that do not affect performance parameters. Specific or general relief is not appropriate for this activity. If the licensee cannot demonstrate that the packing adjustment does not adversely affect performance parameters, the Code requirements must be met for post-maintenance testing. Therefore, the licensee must consider this issue for each valve individually.

Basis for Recommendation

The NRC staff would not require a licensee to shut down a plant to perform IST unless the licensee has no alternative to ensure that the operational readiness of components is maintained or unless a safety issue exists. The IST requirements do not prohibit or discourage a licensee from making limited adjustments to valve packing to stop a leak that may be adversely affecting the valve or surrounding components. Therefore, the licensee can perform an analysis of the packing adjustment and, upon demonstrating that the adjustment does not adversely affect the stroke time (or leakage rate) such that it would not exceed its limiting value, can make the adjustment without a post-maintenance stroke time measurement (or leakage test). Confirmatory testing must be performed at the first available opportunity when plant conditions allow testing. This guidance applies only to valves that need adjustment during power operation and cannot be fully stroked in the plant operating mode. The guidance does not apply merely as a convenience to the licensee and does not supersede any related guidance associated with GL 89-10. NRC IN 87-40, "Back Seating Valves Routinely to Prevent Packing Leakage," gives information related to backseating valves. Both Westinghouse and General Electric had issued guidance on performing backseating to minimize deformation to valve stems. Backseating is not listed as an example of a maintenance activity in ASME OM Subsection ISTC-3310. The licensee would have to assess the effect of backseating on valve operation and determine whether post-maintenance testing is required. Test results for MOV programs to address GL 89-10 and GL 96-05 may indicate whether backseating of a particular valve affects its stroke time. Any information would need to be included and documented in an evaluation, and the assessments would have to be valve-specific.

4.4.3 Manual Valves

The staff has received questions about the requirements for including manual valves in the IST program. The Code includes manual valves that meet the scope requirements of 10 CFR 50.55a. To comply with the OM Code, manual valves must be included in the IST Program and tested in accordance with applicable requirements of Subsection ISTC if they are required to perform a specific function in shutting down a reactor to the safe shutdown condition, in maintaining the safe shutdown condition, or in mitigating the consequences of an accident. Applicable tests could include exercising, leak testing, and/or position indication verification, at the frequency specified in the Code. For some valves, no tests are specified depending on their active/passive classification and their performance and design attributes. For example, a passive manual valve with no position indication has no tests specified. However, it must still be considered and listed in the IST Program.

4.4.3.1   Manual Valve Exercise Interval

The rule published in *Federal Register* (67 FR 60520) on September 26, 2002, 10 CFR 50.55a(b)(3)(vi) included a modification to the ASME OM Code 1998. That modification specified the manual valve exercise interval as 2 years, rather than 5 years as specified in the 1999 and 2000 Addenda to the ASME OM Code. The 1998 Edition and earlier versions of the ASME Code specified an exercise interval of 3 months for manual valves within the scope of the Code. The 1999 Addenda to the ASME OM Code revised Subsection ISTC- 3540 to extend the exercise frequency for manual valves to 5 years; however, the NRC staff did not agree that there is sufficient justification to extend the exercise interval for manual valves to 5 years.

The staff's review of licensees' IST programs indicated that manual valves are exercised every 3 months except in instances where it is impractical to operate valves during unit operation. In such instances, the valves are exercised when the unit is in a cold shutdown or refueling outage condition, and the exercise frequency cannot exceed 2 years. Therefore, a 2-year interval is justified for exercising manual valves because the available manual valve exercise data supports a 2-year interval. Nonetheless, licensees are not prohibited from exercising manual valves more frequently than every 2 years. In the 2006 Addenda to the ASME OM Code revised ISTC-3540 to change the exercise frequency for manual valves to 2 years.

4.4.4 Pressure Isolation Valves

Pressure isolation valves (PIVs) are defined as two normally closed valves in series that isolate the reactor coolant system (RCS) from an attached low-pressure system. PIVs are located at all RCS/low-pressure system interfaces. As such, PIVs are located within the reactor coolant pressure boundary (RCPB), which is defined in 10 CFR 50.2, "Definitions."

"Event V" PIVs are defined as two check valves in series at an RCS/low-pressure system interface, which may result in a LOCA that bypasses containment if they fail. The "Event V" PIVs comprise a subset of PIVs. "Event V" refers to the scenario described for this event in the "Reactor Safety Study" (WASH-1400).

On April 20, 1981, the NRC issued Orders to 32 PWRs and 2 BWRs, which required the specified licensees to conduct leak rate testing of their PIVs, based on plant-specific NRC-supplied lists of PIVs, and required the licensees to modify their TS accordingly. These Orders are known as the "Event V Orders," and the valves listed therein are the "Event V" PIVs.

Currently, the majority of operating plants operate using NUREG 1430 – 1434, Standard Technical Specifications (STS) that do not contain a listing of PIVs or Event V PIVs. Therefore, the staff recommends that licensees should include a listing of PIVs (including Event V PIVs) in their 10-year IST programs to document IST testing requirements for each PIV. Licensees should also review their testing procedures to ensure that the PIVs are individually leak rate tested. (This position supersedes Position 4 of GL 89-04, because the improved STS no longer contain PIV listings.)

4.4.4.1   PIV Discussion in Generic Letter 87-06

GL 87-06 supersedes Position 4 of GL 89-04, because the STS do not contain PIV listings. The staff used licensees' responses to GL 87-06 as input for the resolution of Generic Issue 105, "Interfacing Systems LOCAs at Light-Water Reactors," which was evaluated by the NRC's Office of Nuclear Regulatory Research. The results of studies of interfacing system LOCAs are provided in NUREG/CR-5124, "Interfacing Systems LOCA: Boiling-Water Reactors," and NUREG/CR-5102, "Interfacing Systems LOCA: Pressurized-Water Reactors." Generic Issue 105, which included the issue discussed in GL 87-06, was closed by memorandum from E. Beckjord to J. Taylor, "Technical Resolution of Generic Issue 105 (GI-105), 'Interfacing Systems Loss-of-Coolant Accident (ISLOCA) in LWRs,'" dated June 3, 1993. Pressure isolation valves need to be included in and tested by the IST programs if they are not included as part of a licensee's technical specifications.

4.4.4.2   Leak Rate Testing of PIVs

The leak rate testing specified in a plant's TS must meet the intent of Subsection ISTC-3600. A licensee must ensure that each PIV is individually leak tested (or that the measured leakage is adjusted) in accordance with the differential pressure requirements of the OM Code. If the TS are not sufficiently detailed to ensure individual valve leak testing, the licensee is responsible to ensure that the test procedures are themselves adequate for valves and valve combination leak testing.

NRC Recommendation

A licensee may consider the leakage testing performed to meet TS requirements to also meet IST requirements if the intent of the OM Code is met (e.g., leakage limits are established, corrective actions are taken as required, and valves are individually leak tested). However, a licensee must ensure that the test differential pressure specified in the TS, if applicable, is equivalent to the function maximum pressure differential, or that the measured leakage is adjusted to the function maximum pressure differential in accordance with the formula in Subsection ISTC-3630 of the OM Code.

Basis for Recommendation

Increasing pressure usually improves the seating of a valve. The Code allows that when leak testing those types of valves in which the service pressure will tend to diminish the overall leakage channel opening, as by pressing the disk into or onto the seat with greater force, the test differential pressure may be lower than the function maximum differential pressure.

The resulting leakage is to be adjusted according to the following formula from the OM Code:

$$\frac{L\,(maximum)}{L(test)} = \sqrt[2]{\frac{dP\,(maximum)}{dP(test)}}$$

where

L	= leakage
dP	= differential pressure

While the NRC staff has accepted other aspects of the TS, the licensee must ensure that any testing requirements that are not specifically detailed in the TS are, nonetheless, imposed on the pressure isolation valves to comply with the OM Code leakage testing requirements. Generally, the same test will be used to meet both TS and IST requirements. The major difference between TS and IST requirements are related to the acceptance criteria specified in some TS between a nominal leakage limit and the upper leakage limit. (If allowed by TS, the upper leakage limit is considered acceptable as the acceptance criteria for IST.)

If the list of PIVs is removed from the TS, the leakage testing must be described in detail in the SAR or must be identified as in accordance with the requirements of the ASME OM Code.

4.4.5 Containment Isolation Valves That Have Other Leak-Tight Safety Functions

Valves that function as containment isolation valves may have additional safety functions (i.e., other than isolation), such as pressure isolation, train separation, or preventing diversion of flow. The leakage testing for Appendix J may not adequately test these additional functions based on the pressure or fluid medium. For such valves, the requirements of both Appendix J and Subsection ISTC-3600 apply.

4.4.6 Testing Individual Scram Valves for Control Rods in Boiling-Water Reactors

BWRs are equipped with bottom-entry hydraulically driven control rod drive mechanisms with high-pressure water providing the hydraulic power. Each control rod is operated by a hydraulic control unit (HCU), which consists of valves and an accumulator. The HCU is supplied charging and cooling water from the control rod drive pumps, and the control rod operating cylinder exhausts to the scram discharge volume. Various valves in the control rod drive system perform an active function in scramming the control rods to rapidly shut down the reactor.

The NRC staff believes that those ASME Code-Class valves that must change position to provide the scram function should be included in the IST program and should be tested in accordance with the requirements of Subsection ISTC except where relief has been granted in a safety evaluation report. Bi-directional exercise testing of check valves is required by the 1996 Addenda to the ASME Code (and later editions and addenda).

The control rod drive system valves that perform an active safety function in scramming the reactor are the scram discharge volume vent and drain valves, scram inlet and outlet valves, scram discharge header check valves, charging water header check valves, and cooling water header check valves. With the exception of the scram discharge volume vent and drain valves, exercising the other valves quarterly during power operations could result in rapid insertion of one or more control rods. If practical, licensees should test control rod drive system valves at the Code-specified frequency. However, for those control rod drive system valves for which testing could result in rapid insertion of one or more control rods, the rod scram test frequency identified in the facility's TS may be used as the valve testing frequency to minimize rapid reactivity transients and wear of the control rod drive mechanisms. This alternative test frequency which is deviation from the Code requirement should be clearly stated and documented in the IST program document and this alternative or relief require NRC approval.

Industry experience has shown that normal control rod motion may verify the cooling water header check valve moving to its safety function position. This can be demonstrated because rod motion may not occur if this check valve were to fail in the open position. If this test method is used at the Code-required frequency, the licensee should clearly explain in the IST program document that this is how these valves are being verified to close quarterly.

Closure verification of the charging water header check valves requires that the control rod drive pumps must be stopped to depressurize the charging water header. This test should not be performed during power operation because stopping the pumps results in a loss of cooling water to all control rod drive mechanisms, and seal damage could result. Additionally, this test cannot be performed during each cold shutdown because the control rod drive pumps supply seal water to the reactor recirculation pumps, and one of the recirculation pumps is usually kept running. Therefore, the HCU accumulator pressure decay test, as identified in the facility's TS may be used as the charging water header check valve alternative testing frequency for the reasons stated above. If this test is not addressed in the licensee's TS, this closure verification should be performed at least during each refueling outage, and this alternative test frequency which is deviation from the Code requirement should be specifically addressed in the IST program document and this alternative or relief require NRC approval.

The scram inlet and outlet valves are power-operated valves that full-stroke in milliseconds and are not equipped with indications for both positions; therefore, it may be impractical to measure their full-stroke time as required by the Code. Verifying that the associated control rod meets the scram insertion time limits defined in the plant's TS can be an acceptable alternative method of detecting degradation of these valves. Also, it may be impractical and unnecessary to trend the stroke times of these valves because they are indirectly stroke timed, and no meaningful correlation may be drawn between the scram time and valve stroke time. Furthermore, conservative limits are placed on the control rod scram insertion times. If the above test is used to verify the operability of scram inlet and outlet valves, it should be specifically documented in the IST program document as discussed above.

4.4.7 Use of Appendix J, Option B, in Conjunction with ISTC Exercising Tests

In the ASME OM Code, Subsection ISTC-3522 requires licensees to exercise Category C valves every 3 months. ISTC-3620 also requires licensees to seat leak test Category A valves (containment isolation valves) in accordance with Appendix J to 10 CFR Part 50. Specifically, Option B of Appendix J allows a variable seat leak testing frequency, based on component performance, and allows test intervals for valves with acceptable performance to be extended to once every three refueling outages. Therefore, for Category A/C valves, the Code requires two independent tests, including an exercising test and a seat leakage rate test.

The Code recognizes that when more than one distinguishing category characteristic applies, all requirements for each of the individual categories apply, although duplication or repetition of common testing requirements is not necessary. Therefore, a seat leak rate test is one acceptable method to verify the closure portion of an exercise test.

Appendix II, "Check Valve Condition Monitoring Program," allows an alternative to the exercising testing requirements in the OMa-1996 Addenda. The OMa-1996 Addenda included two significant changes to IST of check valves to (1) correct certain anomalies in the way check valves are currently being exercised, and (2) codify a process for monitoring the valve's operating condition and performance. This integral two-part improvement to the Code provides interrelated requirements. ASME modified Subsection ISTC 4.5.2, "Exercising Requirements," and Subsection ISTC 4.5.4, "Valve Obturator Movement," to require bidirectional testing to improve on the detection of valve degradation and failure. The related change to Subsection ISTC 4.5.5, Condition Monitoring Program, allowed the use of a codified condition monitoring process as an alternative to the exercising and testing requirements of Subsections ISTC 4.5.1 - 4.5.4. The similar requirements are incorporated in the ASME OM Code, Subsections ISTC-3520, "Exercising Requirements," and ISTC-3530, "Valve Obturator Movements," to continue bidirectional check valve testing. The condition monitoring process, defined in Appendix II, "Check Valve Condition Monitoring Program," gives licensees certain IST flexibility in establishing the types of test, examination, and preventive maintenance activities and their associated intervals, when justified based on the valve's performance and operating condition.

NRC Recommendation

The use of the alternative Appendix II Condition Monitoring Program, with the regulatory modifications, provides the licensee with knowledge of the valve's operating condition, informed and verified expectations of the valve's performance over extended intervals, and a process to justify prudent IST interval extensions to reduce the burden of unnecessary IST. Therefore, the staff recommends that licensees implement the condition monitoring program to justify extending the exercise test interval to the leak test frequencies specified in Option B of Appendix J.

Basis for Recommendation

The use of the Appendix-II, "Condition Monitoring Program," as incorporated by reference in 10 CFR 50.55a, provides licensees with knowledge of the valve's operating condition, monitors and verifies valve performance over extended intervals, and provides a process to justify prudent IST interval extensions to reduce the burden of unnecessary IST.

4.4.8 Pyrotechnic-Actuated Valves in New Reactors

Some designs for new nuclear power plants include pyrotechnic-actuated (squib) valves that have more safety significance than squib valves in currently operating nuclear power plants. In addition, squib valves for new reactors might have different designs and be much larger than squib valves used in current plants. Paragraph ISTC-5260 in the ASME OM Code as currently incorporated by reference in 10 CFR 50.55a specifies that at least 20 percent of the charges in explosively activated valves shall be fired and replaced at least once every 2 years. If a charge fails to fire, the ASME OM Code specifies that all charges with the same batch number shall be removed, discarded, and replaced with charges from a different batch. The NRC staff considers these provisions for IST surveillance of squib valves in the ASME OM Code to be insufficient for the design and application of squib valves in some new reactors. At this time, reactor vendors for new nuclear power plants have not completed the design and qualification of squib valves to be used in their new reactors. The NRC staff is monitoring the design and qualification process for squib valves to be used in new reactors by the applicable reactor vendors. The staff is also participating in international efforts to provide improved design, qualification, and testing for squib valves to be used in new reactors.

Nuclear power plant applicants and licensees for new reactors must incorporate the lessons learned from the design and qualification process in the development of IST surveillance activities for squib valves. For example, in addition to test firing sample of squib valve charges specified in the ASME OM Code, licensees of new nuclear power plants should address the following aspects in providing reasonable assurance of the operational readiness of squib valves: (1) verification of the structural integrity of external and internal parts of the actuator and valve; (2) identification and removal of foreign material, fluid and corrosion within the actuator and valve that might interfere with the operation of the actuator or valve; and (3) confirmation of the capability of the pyrotechnic charge in the actuator to provide the necessary motive force to operate the valve under design-basis conditions without damage to the valve body or connected piping.

ASME has prepared updated PST and IST testing and surveillance requirements for squib valves to be used in nuclear power plants licensed after January 1, 2000, in the published 2012 edition of the ASME OM Code. Therefore, nuclear power plants licensed under 10 CFR Part 52 will need to evaluate the applicability of the new squib valve surveillance requirements when implementing the ASME OM Code incorporated by reference in 10 CFR 50.55a 12 months before fuel loading.

To supplement ASME OM Code provisions for squib valves prior to the 2012 Edition, the NRC specified license conditions for PST and IST surveillance of squib valves when issuing the COLs for Vogtle Units 3 and 4 and VC Summer Units 2 and 3. The license condition includes the following requirements:

Before initial fuel load, the licensee shall implement a surveillance program for specific explosively actuated valves (squib valves) that includes the following provisions in addition to the requirements specified in the ASME OM Code as incorporated by reference in 10 CFR 50.55a.

a. Preservice Testing

All explosively actuated valves shall be preservice tested by verifying the operational readiness of the actuation logic and associated electrical circuits for each explosively actuated valve with its pyrotechnic charge removed from the valve. This must include confirmation that sufficient electrical parameters (voltage, current, and resistance) are available at the explosively actuated valve from each circuit that is relied upon to actuate the valve. In addition, a sample of at least 20 percent of the pyrotechnic charges in all explosively actuated valves shall be tested in the valve or a qualified test fixture to confirm the capability of each sampled pyrotechnic charge to provide the necessary motive force to operate the valve to perform its intended function without damage to the valve body or connected piping. The sampling must select at least one explosively actuated valve from each redundant safety train. Corrective action shall be taken to resolve any deficiencies identified in the operational readiness of the actuation logic or associated electrical circuits, or the capability of a pyrotechnic charge. If a charge fails to fire or its capability is not confirmed, all charges with the same batch number shall be removed, discarded, and replaced with charges from a different batch number that has demonstrated successful 20 percent sampling of the charges.

b. Operational Surveillance

Explosively actuated valves shall be subject to the following surveillance activities after commencing plant operation:

(1) At least once every 2 years, each explosively actuated valve shall undergo visual external examination and remote internal examination (including evaluation and removal of fluids or contaminants that may interfere with operation of the valve) to verify the operational readiness of the valve and its actuator. This examination shall also verify the

appropriate position of the internal actuating mechanism and proper operation of remote position indicators. Corrective action shall be taken to resolve any deficiencies identified during the examination with post-maintenance testing conducted that satisfies the PST requirements.

(2) At least once every 10 years, each explosively actuated valve shall be disassembled for internal examination of the valve and actuator to verify the operational readiness of the valve assembly and the integrity of individual components and to remove any foreign material, fluid, or corrosion. The examination schedule shall provide for each valve design used for explosively actuated valves at the facility to be included among the explosively actuated valves to be disassembled and examined every 2 years. Corrective action shall be taken to resolve any deficiencies identified during the examination with post-maintenance testing conducted that satisfies the PST requirements.

(3) For explosively actuated valves selected for test sampling every 2 years in accordance with the ASME OM Code, the operational readiness of the actuation logic and associated electrical circuits shall be verified for each sampled explosively actuated valve following removal of its charge. This must include confirmation that sufficient electrical parameters (voltage, current, resistance) are available for each valve actuation circuit. Corrective action shall be taken to resolve any deficiencies identified in the actuation logic or associated electrical circuits.

 (4) For explosively actuated valves selected for test sampling every 2 years in accordance with the ASME OM Code, the sampling must select at least one explosively actuated valve from each redundant safety train. Each sampled pyrotechnic charge shall be tested in the valve or a qualified test fixture to confirm the capability of the charge to provide the necessary motive force to operate the valve to perform its intended function without damage to the valve body or connected piping. Corrective action shall be taken to resolve any deficiencies identified in the capability of a pyrotechnic charge in accordance with the PST requirements.

This license condition shall expire upon (1) incorporation of the above surveillance provisions for explosively actuated valves into the facility's inservice testing program, or (2) incorporation of inservice testing requirements for explosively actuated valves in new reactors (i.e., plants receiving a construction permit, or combined license for construction and operation, after January 1, 2000) to be specified in a future edition of the ASME OM Code as incorporated by reference in 10 CFR 50.55a, including any conditions imposed by the NRC, into the facility's inservice testing program.

This license condition supplements the current requirements in the ASME OM Code for explosively actuated valves, and sets forth requirements for both preservice testing and operational surveillance, as well as any necessary corrective action. The license condition will expire when either (1) the license condition is incorporated into the plant-specific IST program; or (2) the updated ASME OM Code requirements for squib valves in new reactors, as accepted by the NRC in 10 CFR 50.55a, are incorporated into the plant-specific IST program. For the purpose of satisfying the license condition, the licensee retains the option of including in its IST program either the requirements stated in this condition, or including updated ASME Code requirements.

Applicants and licensees for new reactors will need to address the PST and IST provisions for squib valves consistent with the applicable regulatory requirements, license conditions, and FSAR provisions.

5. SUPPLEMENTAL GUIDANCE
ON INSERVICE TESTING OF PUMPS

5.1   General Pump Inservice Testing Issues

In 1995, OM Code Subsection ISTB introduced a new approach to pump testing, in which pumps are divided into two basic groups with an enhanced baseline or preservice and three periodic tests (i.e., Group A, Group B, and Comprehensive). This modified pump testing program is commonly referred to as the Comprehensive Pump Test (CPT). The pump grouping criterion of ISTB is based on the way the pumps are operated at the plant.

The CPT allows less-rigorous pump testing to be performed for certain pumps on a quarterly frequency while requiring a pump test to be performed with more accurate pressure and differential pressure instrumentation every 2 years at ☐20 percent of pump design flow. The CPT was developed with the knowledge that some pumps, such as containment spray pumps, cannot be tested at the required high flow rates because of system design limitations. Subsection ISTB-3300(e)(1) of the OM Code requires licensees to establish reference values within ☐20 percent of the design flow for the CPT.

5.1.1   Categories of Pumps for Inservice Testing

The ASME OM Code (2004 Edition) requires that all pumps that the licensee identifies as part of an IST program must be categorized as either Group A or Group B pumps. Subsection ISTB-2000 defines Group A as "pumps that are operated continuously or routinely during normal operation, cold shutdown, or refueling operation." By contrast, the Code defines Group B pumps as "pumps in standby systems that are not operated routinely except for testing."

5.1.2   Testing Requirements and Frequency of Inservice Tests

The ASME OM Code identifies four types of tests, including Preservice, Group A, Group B, and Comprehensive tests. All pumps receive a Preservice test followed on a quarterly basis by the test associated with the pump category (Group A test for Group A pumps, and Group B test for Group B pumps), and at least once every 2 years by a Comprehensive test. A Comprehensive test may also be substituted for a Group A or Group B test. Similarly, a Group A test may be substituted for a Group B test, and a Preservice test may be substituted for any inservice test.

ASME OM Code, ISTB-3410, "Pumps in Regular Use," states -

> Group A pumps that are operated more frequently than 3 months need not be run or stopped for a special test provided the plant records show the pump was operated at least once every 3 months at reference conditions, and the quantities specified were determined, recorded, and analyzed per Article ISTB-6000.

ASME OM Code, ISTB-3420, "Pumps in Systems Out of Service," states -

> For a pump in a system declared inoperable or not required to be operable, the test schedule need not be followed. Within 3 months before the system is placed in an operable status, the pump shall be tested and the test schedule followed in accordance

with the requirements of this Subsection. Pumps which can only be tested during plant operation shall be tested within 1 week following plant startup.

ASME OM Code, ISTB-3430, "Pumps Lacking Required Fluid Inventory"

Group B pumps lacking required fluid inventory (e.g., pumps in dry sumps) shall receive a comprehensive test at least once every 2 years except as provided in ISTB-3420. The required fluid inventory shall be provided during this test. A group B test is not required.

NRC Recommendation

In 1995, OM Code Subsection ISTB introduced the new CPT, which allows licensees to perform less-rigorous pump testing for certain pumps on a quarterly frequency, while requiring licensees to perform a pump test with more accurate pressure/differential pressure instrumentation every 2 years at □20 percent of pump design flow. This section also discusses previously issued guidance and experience.

5.2 Use of Variable Reference Values for Flow Rate and Differential Pressure During Pump Testing

Some system designs do not allow for testing at a single reference point or a set of reference points. In such cases, it may be necessary to plot pump curves to use as the basis for variable reference points. Consequently, the ASME Code Committee introduced Code Case OMN-9, Revision 0, "Use of Pump Curves for Testing," which the NRC staff subsequently included in RG 1.192. That regulatory guide lists the OM Code Cases that the NRC staff finds acceptable for licensees to implement in their IST programs for light-water-cooled nuclear power plants. In particular, the staff accepted Code Case OMN-9, with the condition that (1) when the repair, replacement, or routine servicing of a pump may have affected a reference curve, the licensee must determine a new reference curve, or reconfirm an existing reference curve, in accordance with Section 3 of Code Case OMN-9; and (2) if it is necessary or desirable, for some reason other than that stated in Section 4 of Code Case OMN-9, to establish an additional reference curve or set of curves, the licensee must determine the new curves in accordance with Section 3 of Code Case OMN-9. The use of OMN-9 requires relief because OMN-9 is only applicable to the ASME OM Code-1990 through ASME OMb Code 1992. It is not applicable to 1995 or later Code editions. In ASME OMb-2006 addenda, the ASME OM Code committee developed a new Code Case OMN-16, "Use of a Pump Curve for Testing," to replace Code Case OMN-9. The Code Case OMN-16 incorporates all the NRC conditions imposed for the use of Code Case OMN-9. Therefore, once the Code Case OMN-16 is endorsed in RG 1.192, no relief request is need to use OMN-16.

5.2.1 Reference Values

Licensees shall determine reference values from the results of Preservice testing or the first inservice test. The resultant reference values shall be at points of operation that are readily duplicated during subsequent tests, and the licensee shall compare all subsequent test results to the initial reference values or the new reference values established in accordance with the Code. Licensees shall only establish reference values when the pump is known to be operating acceptably. If the particular parameter being measured or determined can be significantly influenced by other related conditions, these conditions shall be analyzed.

5.2.2   Reference Curves

If the establishment of specific reference values is impractical for a centrifugal or vertical line shaft pump, the licensee may establish reference curves. In so doing, the licensee shall determine the reference curves from the data measured during Preservice testing or the first inservice test. In addition, the licensee shall establish a reference curve from a minimum of five data points for each 20 percent of the maximum pump curve range, and the range of the reference curve shall be sufficient to bound the points of operation that are expected during subsequent tests. The licensee shall then compare all results to the initial reference curves or the new values established in accordance with Sections 5.2.3 and 5.2.4, below. In addition, the licensee shall only establish reference curves when the pump is known to be operating acceptably. If vibration is relatively unaffected by changing differential pressure or flow over the reference curve range, the licensee may use a single reference value as the test quantity, provided it is at the minimum of the measured data. By contrast, if the licensee uses reference curves, the record of the test shall document and justify the reasons for doing so and the suitability of the methods used to develop the reference curves and acceptance criteria. (See Subsection ISTB-9000.)

5.2.3   Effect of Pump Replacement, Repair, and Maintenance on Reference Values or Reference Curves

When the repair, replacement, or routine servicing of the pump may have affected a reference value, a set of reference values, or a reference curve, the licensee shall determine a new reference value, set of reference values, or a new reference curve or reconfirm the previous value (or curve) by an inservice test run before declaring the pump operable. The licensee shall then identify any deviation between the previous and new set of reference values (or reference curves), and the record of the tests shall document the verification that the new values (or curves) represent acceptable pump operations. (See ISTB-9000)

5.2.4   Establishment of Additional Sets of Reference Values or Reference Curves

If it is necessary or desirable, for some reason other than discussed above, to establish an additional set of reference values or curves, the licensee shall run an inservice test under the conditions of an existing set of reference values, or within the range of existing reference curves, and shall analyze the results. If the operation is acceptable in accordance with Section 7 of Code Case OMN-9 (Section 16-6200 of Code Case OMN-16), a second test run under the new reference conditions shall follow as soon as practicable, and the results of this test shall establish the additional set of reference values or reference curves. Whenever a licensee establishes an additional set of reference values or reference curves, the record of the tests shall document and justify the rationale for doing so. (See ISTB-9000.)

NRC Recommendation

The NRC accepts the use of pump curves for reference values of flow rate and differential pressure if the licensee clearly demonstrates, in a relief request, that it would be impractical to establish a fixed set of reference values. A relief request must include a description of the methodology to be used in evaluating these pumps. To obtain approval for a proposed method of evaluating these pump parameters to detect hydraulic degradation and determine pump operability, the licensee must demonstrate that the acceptance criteria are equivalent to the Code requirements as specified under test acceptance criteria in ISTB Table ISTB-5121-1 or ISTB-5221-1 (depending on pump type), for allowable ranges using reference values and curves.

To use this test method, the licensee must plot a valid pump characteristic curve from empirical data or obtain one from the pump manufacturer and verify it with measurements taken when the pump was known to be in good operating condition. Additional guidance is given in Sections 5.2.2, 5.2.3, and 5.2.4 above; the OM Code; and Code Case OMN-9, including RG 1.192.

The use of OMN-9 requires relief because OMN-9 is only applicable to the ASME OM Code-1990 through ASME OMb Code 1992. It is not applicable to 1995 or later Code editions. Code Case OMN-16 may be used once endorsed in RG 1.192.

Basis for Recommendation

Where it is not practical to return to the same flow configuration for each subsequent inservice pump test, the licensee must establish a method for evaluating the operational readiness of pumps in variable flow systems. This may be the case for service water or component cooling water systems and other systems where temperature or flow is controlled at a variety of locations. During quarterly pump testing, the licensee may not be able to manually control each of these local stations and duplicate the overall system reference conditions, as required by the Code.

Using the manufacturer's pump-specific curves for flow and differential pressure, the licensee may be able to evaluate the pump in as-found system conditions. In implementing this guidance, the licensee would confirm these values by performing in situ testing. Another method would be to plot pump curves over the range of conditions expected during the system's normal operation. It is also important to develop a method of evaluating pump vibration measurements taken with the pump operating over the range of possible as-found conditions, since this is a variable pump parameter. By evaluating these measurements of pump vibration, the licensee will ensure that a pump that is severely degraded, either hydraulically or mechanically, is declared inoperable and appropriate action is taken to address the degradation.

5.3   Allowable Variance from Reference Points and Fixed-Resistance Systems

Certain designs do not allow for the licensee to set the flow at an exact value because of limitations in the instruments and controls for maintaining steady flow. The characteristics of piping systems in other designs do not allow for the licensee to adjust the flow to exact values. The Code does not allow for variance from a fixed reference value, stating only that "the resistance of the system shall be varied until either the measured differential pressure or the measured flow rate equals the corresponding reference value." Licensees have requested relief to establish a range of values similar to using a pump curve, but with a very narrow band. For example, one licensee proposed to use a reference curve with the tolerance around the selected value of flow to be ☐2 percent. Plant implementing procedures may instruct operators to set the flow to 1,500 gallons per minute (gpm). When this step is performed, the operator would attempt to set the flow as close as possible to 1,500 gpm and maintain it steadily at approximately 1,500 gpm.

NRC Recommendation

The staff has determined that, if the design does not allow for establishing and maintaining flow at an exact value, achieving a steady flow rate or differential pressure at approximately the set value does not require relief for establishing pump curves. The allowed tolerance for setting the fixed parameter must be established for each case individually, including the accuracy of the instrument and the precision of its display. This will necessitate verification of the effect of precision on accuracy as considered in the design of the instrument gauge. For Group A and Group B tests, a total tolerance of ☐2 percent of the reference value (including instrument accuracy) is allowed without prior NRC approval; for Preservice and Comprehensive tests, the allowable total tolerance is ☐1/2 percent (including instrument accuracy) for pressure and differential pressure, ☐ 2 percent (including instrument accuracy) for flow. For a tolerance greater than the allowed percent (which may be necessary depending on the precision of the instrument), the licensee may make a corresponding adjustment to acceptance criteria to compensate for the uncertainty, or may perform and document an evaluation to justify a greater tolerance. In using this guidance, the IST program document or implementing procedures must document the variance and the method for establishing it.

The intent is that the variance in the reference value setting may be ☐2 percent for Group A and Group B tests and ☐1/2 percent for Preservice and Comprehensive tests, ☐1/2 for pressure and differential pressure, and ☐2 percent for flow without requiring relief. Nonetheless, any variance in the setting will have an impact on test results.

Basis for Recommendation

The OM Code does not address the likelihood that it may not be possible to control a flow rate or differential pressure to an exact value. When the Code specifies that the system resistance must be varied until either the flow or differential pressure equals the corresponding reference value, it does not intend the set value to have an acceptable range as stated in the ISTB test acceptance criteria, including ISTB Tables ISTB-5121-1 and ISTB-5221-1. The acceptance criteria apply only to the parameter being determined after the resistance is varied. Licensees should recognize that the reference value for certain pumps can only be achieved within a specified tolerance. Licensees may set the repeatable parameter as close as possible to the reference value during each test, rather than treating any variance in the value with a pump curve. If, upon establishing trends in data, the licensee determines that the parameter varies such that the readings are outside the accuracy of the instrument, the licensee may need to establish pump curves and propose an alternative to the Code requirements for the applicable pumps. (See Section 5.2)

Subsection ISTB-3500 specifies the requirements for instrument fluctuations and describes the basis for allowing a variance from the reference value of ☐2 percent for Group A and Group B tests, and ☐1/2 percent for pressure and differential pressure, ☐2 percent for flow for Preservice and Comprehensive tests. In addition, Subsection ISTB-3500 allows the use of symmetrical damping devices or averaging techniques to reduce instrument fluctuations to within ☐2 percent or ☐1/2 percent (as applicable) of the observed reading for values specified in the implementing procedures. Greater variances must be justified and acceptance criteria adjusted as necessary.

If an analog gauge is used, the required accuracy is percent full scale. For a digital gauge, the required accuracy is over the calibrated range. For a combination of gauges, the required accuracy is loop accuracy.

The ASME is currently in the process of developing a Code change as follows to address this issue:

The allowed tolerance for setting the fixed parameter must be established for each case individually, including evaluation of throttling capability. Licensees should consider improvements in throttling methods where system control is especially poor. A total throttling tolerance of + 2 / -1 percent of the flow rate reference value is considered as meeting the requirements of the code sections.

For a tolerance greater than + 2 / -1 percent of the flow rate reference value, a corresponding adjustment to acceptance criteria shall be made to compensate for the uncertainty, or an evaluation would be performed and documented justifying a greater tolerance. The variance and the method for establishing the variance must be documented in the IST program documents or implementing procedures.

The basis for the Code change is as follows:

The ASME OM Code does not address the possibility that a flow rate or differential pressure may not be controllable to an exact value. When the Code specifies that the system resistance be varied until either the flow or differential pressure equals the corresponding reference value, it does not literally intend that the "set value" be precisely attained without any fluctuations. Licensees recognize that the reference value for certain pumps can only be achieved within a specified tolerance. Licensees shall attempt to set the repeatable parameter as close as possible to the reference value during each test.

The basis for allowing a variance of + 2 / -1 percent from the flow rate reference value deals with instrument fluctuations and system stability issues. The Code allows symmetrical damping devices or averaging techniques to be used to reduce instrument fluctuations to within 2 percent of the observed reading for values specified in the implementing procedures. Greater variances must be justified and acceptance criteria adjustments made as necessary. The limitation of 1 percent in the negative direction reduces the non-conservative impact on the variable parameter. The total 3 percent allowable variance provides for a reasonable throttling control range while minimizing the impact on trendability of the variable parameter.

Licensees should ensure that performance trending of pumps is capable of detecting degradation as early as possible. Larger variances in the reference parameter will induce scatter in the variable parameter data. Techniques such as data normalization, where recorded test data is corrected by the known pressure to flow relationship, should be used when necessary to provide for accurate short term trending.

5.4   Monitoring Pump Vibration in Accordance with ISTB

OM Subsection ISTB allows licensees to monitor pump vibration in units of either pump displacement (peak-to-peak) or pump velocity (peak), and includes acceptance criteria for both units of measurement. As specified in OM Table ISTB-3000-1, the measurement of pump vibration is required for Preservice, Group A, and Comprehensive tests. However, the Code does not require vibration measurements for Group B tests. The staff has determined that if the licensee uses OM Subsection ISTB as the basis for monitoring vibration in the IST program, the program must include all of the requirements for such monitoring. Licensees may update their programs in accordance with this position without further relief if they meet all related requirements for monitoring vibration in Subsections ISTB-3540, ISTB-5000, and ISTB-6000. However, Commission approval to use a later Code edition or addenda is still required pursuant to 10 CFR 50.55a(f)(4)(iv). See Sections 1.1 and 2.1 for further guidance.

In following this guidance, the frequency response range of the instrumentation must be as specified in Subsection ISTB-3510(e) for both low-speed and high-speed pumps, unless the licensee demonstrates that the information gained at the low-frequency response does not apply for the bearing design of the pumps. In that event, the licensee must still provide an acceptable alternative to the required testing. Although the instruments in the low-frequency response ranges may not be widely used, the unavailability of instruments does not constitute sufficient justification for either obtaining relief from the frequency response range requirements of Subsection ISTB, or obtaining approval of an alternative to the requirements.

Basis for Recommendation

As shown in the ASME OM Code, Figure ISTB-5223-1, "Vibration Limits," licensees may choose to use units of velocity, rather than displacement, in measuring vibration in pumps that operate above 600 revolutions per minute (rpm). Such an approach would enable the licensee to more rapidly detect wear in the anti-friction bearing and other types of pump degradation and, thus, to perform repairs in a more timely manner.

Pump bearing degradation results in increased vibration at frequencies of 5 to 100 times the rotational speed of the pump. These high-frequency bearing vibrations may not significantly increase the measured displacement of pump vibration and could go undetected. However, the high-frequency vibrations would significantly increase the measured velocity of pump vibration, which could indicate the need for corrective action before the bearing fails. Because pump bearings vibrate at high frequencies, the measured vibration velocity indicates the mechanical condition of the pumps and reveals pump bearing degradation much more accurately than does measured vibration displacement.

Advantages of measuring vibration velocity, rather than displacement, to monitor the mechanical condition of pumps (with the exception of low-speed pumps) are widely acknowledged in the industry. Many licensees measure pump vibration velocity to detect pump degradation and obtain advance warning of incipient pump bearing failure. Upon obtaining this advance warning, the licensee can plan and prepare for maintenance during scheduled outages instead of suffering losses resulting from unplanned outages to repair failed critical equipment. ASME OM Code, Subsection ISTB includes a set of allowable ranges for inservice pump vibration velocity and measured pump vibration displacement. These ranges are based on an evaluation of empirical data and various acceptance criteria for pump vibration velocity established by U.S. industries, academia, international industry, and foreign agencies.

The minimum frequency response range requirement is established from one-third of the minimum pump shaft rotational speed to at least 1000 Hz in order to encompass all noise contributors that could indicate degradation. Instruments with a frequency response range that meets these requirements for slow-speed pumps may not be widely used. However, the unavailability of instruments, alone, does not constitute adequate justification for obtaining relief or approval of an alternative; however, it may be a significant element in the justification. The NRC has observed that, because of technology advancement and research in the field of instrumentation, vibration measuring transducers meeting the Code requirements can now be procured from various suppliers at reasonable costs. Additionally, frequencies less than running speed may not be indicative of problems for certain types of bearings; however, subharmonic frequencies may be indicative of rotor rub, seal rub, loose seals, or coupling damage. The type of bearings and other subharmonic concerns would typically be discussed in the justification for relief.

ASME OM Code, Subsection ISTB requires licensees to measure vibration in units of either pump displacement (peak-to-peak) or pump velocity (peak). Digital equipment can measure directly in peak units. The 10-year update of the ISI and IST programs reflects the need for licensees to incorporate new technologies that have been incorporated into the codes and standards. The ASME OM Code Committee responded to an inquiry (Interpretation 95-4, File OMI 94-2) explaining that the intent of the ASME OM Code is to allow vibration to be measured in root mean square (rms) and mathematically converted to peak readings. Licensees are cautioned that the Code vibration acceptance criteria are in peak or peak-to-peak units, and the use of rms is not acceptable without a mathematical conversion. To comply with the requirements, licensees that use rms values for recording data must adjust the limits of OM Subsection ISTB, or convert the data to peak values.

Several plants have requested an alternative to the vibration acceptance criteria of Section ISTB for smooth-running pumps, and the NRC has approved such requests. However, licensees with such approval must continue to assess the vibration data and monitor increases that may be indicative of a change. In one reported incident, a pump with very low vibration experienced an increase in vibration levels over three successive tests, although the levels remained below the criteria for smooth-running pumps. Upon investigating the cause of the increase, the licensee determined that the bearing had degraded and required replacement.

5.5   Pump Flow Rate and Differential Pressure Instruments

The NRC has received requests for relief to continue using instruments that do not meet either the range or accuracy requirements of the Code. The Code requires each analog instrument to have a full-scale range that is three times the reference value or less, while each digital instrument must be such that the reference values do not exceed 70 percent of the calibrated range of the instrument. The NRC has accepted Code Case OMN-6 as specified in RG 1.192, which allows each digital instrument to be such that the reference values do not exceed 90 percent of the calibrated range of the instrument. For Group A and Group B pumps, OM Subsection ISTB-3510 requires an accuracy of ☐2 percent of full-scale for analog instruments, ☐2 percent of total loop accuracy for a combination of instruments, or ☐2 percent of reading over the calibrated range for digital instruments. For Preservice and Comprehensive tests, the required instrument accuracy is □1/2 percent for pressure and differential pressure instruments.

5.5.1 Range and Accuracy of Analog Instruments

NRC Recommendation

When the range of a permanently installed analog instrument is greater than three times the reference value, but the accuracy of the instrument is more conservative than that required by the Code, the staff may grant relief when the combination of the range and accuracy yields a reading that is at least equivalent to that achieved using instruments that meet the Code requirements (i.e., up to □6 percent for Group A and B tests, and □1.5 percent for pressure and differential pressure instruments for Preservice and Comprehensive tests). The use of a test gauge (in lieu of a permanent instrument) is acceptable if the reading is at least equivalent to that required by the Code. When using temporary instruments, the staff recommends that the licensee's IST records should include an instrument number for use in tracing each instrument and a calibration data sheet for use in verifying that the instruments are accurately calibrated. The licensee need not obtain relief if the temporary instruments meet the range and accuracy requirements of the Code. If relief is requested, the licensee would typically describe the effect on each group of applicable pumps and would typically discuss adjustment of acceptance limits to account for the inaccuracies.

Basis for Recommendation

Because the IST requirements originally specified an instrument range of 4 times the reference values or less, the permanent instruments in many early-licensed plants do not meet the current requirements of the Code for an instrument range of three times the reference values or less. The NRC does not generally consider instrument installation or replacement an undue burden, and compliance with the instrument requirements in later editions of the Code does not constitute a backfit.

This position applies to the early-licensed plants, but not for the purchase of replacement instruments that can be procured to meet the current requirements of the Code; therefore, for new instrument installations, licensees must meet the accuracy and range requirements (although the Code does not prohibit like-for-like instrumentation for the existing installation).

The licensee is not relieved of its responsibility to make modifications to comply with changes to IST as a result of changes to the Code. Instrument modifications are considered practical in the context of 10 CFR 50.55(a)(f)(4). However, the use of any available instruments that meet the intent of the Code requirements for the actual reading would yield an acceptable level of quality and safety for testing. Licensees are required to submit a relief request in this case.

When the licensee submits a relief request, it should separately address each group of affected pumps if the instruments are permanently installed. By contrast, a general relief request may be acceptable for temporary instrumentation. If the instruments do not meet the intent of the Code requirements, the NRC may require the licensee to adjust acceptance limits to account for the instrument inaccuracies.

Licensees are cautioned that the CPT requires more accurate instruments than those specified in earlier editions of the Code. As a result, licensees must verify that instrument accuracy is appropriate for the type of test being performed (Group A or Group B versus a Comprehensive test). Licensees should also note that previously acceptable instruments may no longer be acceptable when updating to a more recent edition of the Code.

5.5.2 Use of Tank Level to Calculate Flow Rate for Positive Displacement Pumps

The NRC has received requests for relief to use the tank level to calculate the flow rate in a system with a positive displacement pump when the system was not designed with a flow meter in the flow loop.

OM Subsection ISTB-3550 requires licensees to measure the pump flow rate using a rate or quantity meter installed in the pump test circuit. If the meter does not directly indicate the flow rate, the record of the test shall identify the method used to reduce the flow data. In addition, Subsection ISTB-5300(a) requires a 2-minute run time in order to achieve stable pump performance parameters before recording data during the test.

NRC Recommendation

When flow meters are not installed in the flow loop of a system with a positive displacement pump, it is impractical to directly measure flow rate for the pump. The staff has determined that, if the licensee uses the tank level to calculate the flow rate as described in Subsection ISTB-3550, the implementing procedure must include the calculational method and any test conditions needed to achieve the required accuracy. Specifically, the licensee must verify that the reading scale for measuring the tank level and the calculational method yield an accuracy within ⬜2 percent for Group A and B tests, and Preservice and Comprehensive Tests. If the meter does not directly indicate the flow rate, the record of the test shall identify the method used to reduce the flow data.

Basis for Recommendation

The OM Code requires licensees to measure the pump flow rate in order to determine the extent of any pump degradation. A minimum pump run time of 2 minutes is required in order to achieve stable performance parameters before recording data during the test.

Requiring licensees to install a flow meter to measure the flow rate and to guarantee the test tank size, such that the pump flow rate will stabilize in 2 minutes before recording data would be a burden because of the design and installation changes to be made to the existing system. Therefore, compliance with the Code requirements would be a hardship.

The average flow rate is calculated by measuring the change in test tank level over a period of time and converting it to flow rate using the following standard formula:

$$Q \text{ (GPM)} = \coprod \Delta L \text{ (inch)} / \Delta t \text{ (Second)}$$

Where: Q is flow rate

$\coprod$ is a constant which reflects tank dimensions and unit conversions

ΔL is the measured change in level in the tank in time Δt.

Pump discharge pressure will match system pressure up to the shutoff head of the positive displacement pump. Because of the characteristics of a positive displacement pump, there should be virtually no change in pump discharge flow rate as a result of the rising tank level. Therefore, rising tank level will not have an impact on test results. By having approximately the same level in the tank at the beginning of each test, licensees can achieve repeatable results. In addition, the suction would be from a large source at a constant pressure, which will allow pump performance parameters to stabilize quickly. This method would provide reasonable assurance of operational readiness, provided that the licensee measures the test tank level in accordance with the accuracy requirements of OM Table ISTB-3500-1. The implementing procedures should document the calculational method and test conditions required to achieve this accuracy. Therefore, the proposed alternative of using the tank level to calculate the flow rate provides reasonable assurance of operational readiness. Licensees must submit a relief request to implement this proposed alternative.

5.5.3   Use of Tank or Bay Level to Calculate Differential Pressure

The NRC has received requests for relief to use the tank or bay level to calculate differential pressure when a direct measurement of inlet pressure or differential pressure is not available.

NRC Recommendation

When inlet pressure gauges are not installed in the inlet of a vertical line shaft pump, it is impractical to directly measure inlet pressure for use in determining differential pressure for the pump. The staff has determined that, if the licensee uses the bay level to calculate the suction (inlet) pressure as described in Subsection ISTB-3520(b), the implementing procedure must include the calculation. The licensee must also verify that the reading scale for measuring the level and the calculational method yield an accuracy within $\square2$ percent for Group A and B tests, and $\square1/2$ percent for Preservice and Comprehensive tests. If direct measurements are impractical for other types of pumps with suction from a tank, the licensee must apply similar controls. The Code allows the licensee to determine differential pressure by obtaining the information from a differential pressure gauge or differential pressure transmitter, or by determining the difference between the pressure at a point in the inlet pipe and the pressure at a point in the discharge pipe (Subsection ISTB-3520(b)). Therefore, the licensee may implement a calculational method without obtaining relief because the ASME Code allows for the determination of differential pressure from the discharge pressure and the pressure in the pump inlet.

Basis for Recommendation

The method is in accordance with a determination of differential pressure allowed by the Code. Although the inlet pressure is not directly measured, it is "measured" for the purpose of determining the pressure at a point in the inlet. By including the calculation in the implementing procedures, the licensee can determine the differential pressure in a manner that is consistent and repeatable from test to test. This method will yield the information needed for monitoring the hydraulic condition of the applicable pumps without the need to install suction (inlet) pressure gauges, which may not be practical, depending on the design limitations in the inlet of the pump.

5.5.4 Accuracy of the Flow Rate Instrument Loop

As clarified in OM Code Interpretations 95-7 (OM-1990, Subsection ISTB 4.6.1 and Table ISTB 4.6.1.1; OM-1987 with OMa-1988, Part 6, Para. 4.6.11 and Table 1, Instrument Accuracy) and Inquiries IN 91-3 (OM-1987 through OMc-1990, Part 6, Para. 4.6.1.1) and IN 91-037 (ASME B&PV Code Section XI, 1977 Edition Through Later Editions and Addenda through the 1987 Addenda, Table IWP-4110-1, Instrument Accuracy-Flowrate), the accuracy requirements of analog instruments that are used to measure process flow apply only to the reference calibration of the instrument, such as that supplied by the instrument manufacturer, in determining loop accuracy. In determining instrument accuracy, the Code does not explicitly require the licensee to consider physical attributes (such as orifice plate tolerances), tap locations, environmental effects (such as temperature, radiation or humidity), vibration effects (such as seismic) or process effects (such as temperature). However, factors associated with attributes that could affect the measurements include the effects of wear, accumulation of dirt or grease on an annubar flow coefficient, and the reversed installation of a one-direction orifice plate.

NRC Recommendation

The Code requirements for instrument accuracy ensure that the instrument loop accuracy is adequate for monitoring pumps for degrading conditions. The accuracy for analog instruments specified in OM Subsection ISTB-3500 applies only to the calibration of the instruments. The staff recommends that, when test results indicate that conditions in the pump or the test circuit have changed, licensees should consider corrective action for other attributes that could affect the overall loop accuracy of the measurements.

Basis for Recommendation

In ASME Code Interpretation 95-7 and ASME Code Inquiries IN 91-3 and IN 91-037, the ASME Code Committee states that the requirements for the final indication of flow rate on an analog instrument to be within 2 percent of full scale of actual process flow rate applies only to the calibration of the instrument and does not take into account physical attributes, environmental effects, vibration effects, or process effects.

5.6 Operability Limits of Pumps

For details see NRC Inspection Manual Part 9900: Technical Guidance, Appendix-C, "Specific Operability Issues," Section C.8, "TS Operability vs. ASME OM Code Criteria."

5.7 Duration of Tests

Subsections ISTB-5100(a), ISTB-5200(a), and ISTB-5300(a), "Duration of Tests," requires that for measuring parameters as specified in Table ISTB-3000-1, each pump shall be run for at least 2 minutes after pump conditions are stable as the system permits. This duration is applicable to Group A tests and CPTs. The staff recommends, if practicable, that this duration also be applied to Group B pump tests.

Basis for Recommendation

The 2-minute run time is adequate after pump operation becomes stable. This 2-minute run time minimizes overheating of pumps that are tested using the minimum flow recirculation line. The NRC recommends that licensees should minimize the time pumps are operated on the minimum flow recirculation line. (See NRC Bulletin 88-04, "Potential Safety-Related Pump Loss.")

5.8   Adjustments for Instrument Inaccuracies

If the accuracy of plant instrumentation used for IST is not well understood, the test results may not be adequate to meet the licensee s safety analysis, even if they meet the Code requirements. For example, TS or the safety analysis report require a pump to produce 1,000 gpm at 500 pounds per square inch differential (psid), but the IST reference values are 1,000 gpm (fixed) and 550 psid. The low end of the acceptable range for differential pressure from ISTB Table ISTB-5121-1 for Group A and Group B tests (0.90) would be 495 psid, although conservatively set at 500 psid. If this test is also to prove operability of the pump in addition to meeting IST requirements, and the ☐2 percent instrument inaccuracies were taken into account for flow rate and differential pressure, there is the possibility that the pump is putting out less than the required values. In this example, the instrument accuracies would need to be taken into account if they were not already considered when the design parameters were developed.

When pump test procedures are developed, limits in the safety analysis cannot be ignored. The IST requirements are written generally. If specific plant limits are more conservative, to ensure compliance with design-basis assumptions, such limits must be clearly indicated as the "operability" limits and used for acceptance criteria of IST. For example, when obtaining values using instrumentation that meets the accuracy requirements specified for "information only" or for IST, the value as-read would be used. If a licensee is attempting to perform a critical test, more accurate instrumentation may be necessary; however, the value recorded would be the value read if the accuracy of the instrumentation met the specified accuracy. Only when instruments are used that cannot meet the specified accuracy for a test would an adjustment be necessary to meet the Code. Design analyses may not account for instrument accuracy readings; however, when the pump selection is made, the designer generally selects from a catalog of available sizes and chooses one with margin above the analyses numbers.

In the determination of loop accuracy, it is intended that only the instrument manufacturer's reference accuracy be considered. It is not necessary to consider all uncertainties (such as environmental effects, process effects, vibration effects, etc.).

5.9 Pump Testing Using Minimum Flow Return Lines With or Without Flow Measuring Devices

The NRC has received relief requests from licensees requesting approval of pump testing by using minimum flow return lines with or without measuring devices as an alternative to the IST requirements as specified in the Subsection ISTB of the OM Code.

As specified in Section 5.1.2 above, the Code identifies four types of tests, including Preservice, Group A, Group B, and Comprehensive tests. All pumps receive a Preservice test followed on a quarterly basis by the test associated with the pump category (Group A test for Group A pumps, and Group B test for Group B pumps), and at least once every 2 years by a Comprehensive test. A Comprehensive test may also be substituted for a Group A or Group B test. Similarly, a Group A test may be substituted for a Group B test, and a Preservice test may be substituted for any inservice test.

Subsections ISTB-5100(b), ISTB-5200(b), and ISTB-5300(b) of the ASME OM Code allow the use of a bypass test loop for Group B tests, provided that it is designed to meet the pump manufacturer's operating specifications (e.g., flow rate, time limitations) for minimum flow operation. The bypass test loop may be used for Group A or Comprehensive tests, provided that the flow rate through the loop meets the requirements specified in Subsection ISTB-3300.

An inservice pump test requires that the pump parameters shown in Table ISTB-3000-1 must be measured and evaluated to determine pump condition and detect degradation. Pump differential pressure and flow rate are two parameters that are measured and evaluated together to determine pump hydraulic performance.

In cases where only the minimum-flow return line is available for pump testing, regardless of the test interval, the staff's position is that flow instrumentation that meets the requirements of Subsection ISTB-3500 should be installed in the mini-flow return line. Installation of this instrumentation is necessary to provide flow rate measurements during pump testing so that this data can be evaluated with the measured pump differential pressure to monitor for pump hydraulic degradation.

When testing a pump using a minimum flow recirculation line, the guidance provided in NRC Bulletin 88-04 applies. Licensees should review operating conditions to ensure that a pump is not subject to dead head conditions and that the minimum flow line is adequately sized and that operation will not damage the pump.

5.10 Alternative to ASME OM Code Comprehensive Pump Testing Requirements

The NRC has received relief requests from licensees requesting approval of alternatives to the Comprehensive Pump Testing (CPT) requirements specified in Subsection ISTB-3300, "Reference Values," of the OM Code.

The CPT was developed with the knowledge that some pumps, such as containment spray pumps, cannot presently be tested at the required high flow rates because of system design limitations. Consequently, Subsection ISTB-3300(e)(1) requires licensees to establish reference values within ☐20 percent of the design flow for the CPT.

Some designs do not allow a CPT at a pump design flow of ☐20 percent because of the original system design configuration. In such cases, it may be necessary to use the pump's recirculation line for IST; however, recirculation lines are not typically designed to accommodate ☐20 percent of the design flow.

NRC Recommendation

The NRC accepts the use of lower flow (reference values) other than ☐20 percent of the design flow, as specified by OM Subsection ISTB-3300(e)(1) for CPT, if the licensee's relief request clearly demonstrates the impracticality of establishing a reference value within ☐20 percent of the design flow for the CPT.

To obtain approval for a proposed alternative method of performing the CPT with a flow other than as specified in Subsection ISTB-3300(e)(1) and measuring pump parameters to detect hydraulic degradation and determine pump operability, the licensee must demonstrate that the acceptance criteria are equivalent to the CPT Code requirements in Subsection ISTB-3300(e)(1).

To show the impracticality, the licensee should include (as a minimum, but not limited to) the following information in the submitted relief request:

(1) Provide reason(s) for not performing the CPT at the required flow of ☐20 percent of pump design flow.

(2) Specify the maximum flow at which the CPT can be performed.

(3) Provide the estimated cost of any temporary or permanent system modification required to enable a CPT to be performed at ☐20 percent of pump design flow, along with any difficulty associated with implementing the modification.

(4) Provide all details (e.g., temporary modifications of piping, containment sump, etc.), including pump performance curves, if a full-flow test was performed during preservice or service of the plant.

(5) Provide pump performance curves and any other data associated with the pump's shop testing provided by the manufacturer.

(6) Provide the records and history of maintenance and repairs performed on the pump.

(7) Provide any appropriate compensatory actions being proposed to supplement the alternative testing, such as (but not limited to) the following examples:

(a) testing at the best efficiency point (BEP) on a longer interval; BEP is defined as the capacity and head at which the pump efficiency is at its maximum

(b) commitment to perform additional performance monitoring

(c) adjustment of acceptance criteria

(d) continuation of the previous Code testing, including taking overall vibration data quarterly

(e) periodic sampling and analysis of the lube oil.

Additional guidelines are included in NUREG/CP-0152, Volume 4, Proceedings of the Seventh NRC/ASME Symposium on Valve and Pump Testing, entitled "Comprehensive Pump Testing Based on ASME OM Code Requirements and its Alternative and Related Relief Requests."

This guidance requires relief because the Code does not allow for a reference value of flow for CPTs, other than at flow rates within □20 percent of pump design flow. The NRC will review any relief requests on a case-by-case basis.

5.11 Waterleg Pumps

The NRC has received proposed alternatives from licensees of boiling-water reactor (BWRs) for Group A tests for waterleg pumps. Subsection ISTB-3400 and Table ISTB-3400-1 of the OM Code specify that a Group A test must be performed quarterly for Group A pumps. The waterleg pumps are low flow pumps that are required to operate whenever their respective emergency core cooling system (ECCS) trains are in the operable condition. As such, the pumps perform continuous duty on a recirculation line and provide makeup as needed. There is typically no flow instrumentation of the recirculation line, and the flow instrumentation on the main ECCS header is not sufficiently accurate to measure the low flow of the pumps. When requesting an alternative Group A test for a waterleg pump, a licensee should explain how the pump discharge pressure is monitored, the main ECCS header is verified to be full of water, what is the pump vibration monitoring frequency, and any other maintenance or testing activity performed to ensure the pump will continue to meet its intended function.

5.12 Smooth-Running Pumps

Pumps that have very low vibration reference values (less than or equal to 0.05 inch per second) are called smooth-running pumps. A small increase in smooth-running pump vibration during the OM Code-required IST causes the pump to exceed OM Code vibration acceptance criteria, which normally results in unnecessary corrective action. The NRC has authorized alternative vibration acceptance criteria for smooth-running pumps on a case-by-case basis in accordance with 10 CFR 50.55a(a)(3).

Alternative requests for smooth-running pumps should specify a minimum vibration reference value ($\leq$ 0.05 inch per second), and these smooth-running pumps must be included in a predictive maintenance (PdM) program. The importance of the PdM program for smooth-running pumps was demonstrated when a plant using NRC-authorized alternative vibration acceptance criteria noted a bearing failure that was not detected by the IST program, but was detected through enhanced vibration monitoring as part of the plant's PdM program. During IST, corrective action was not required because the measured vibration was below the alert range as specified by the Code. After the pump bearing failed, it became clear that a simple minimum vibration reference value alone is not sufficient to identify degradation of a smooth-running pump. PdM programs normally include bearing temperature trending, oil sampling and analysis, thermographic analysis, and enhanced vibration monitoring. The objective of the PdM program should be to detect and correct problems involving the mechanical condition of the pump before the pump reaches its overall vibration alert limit.

5.13   Vibration-Measuring Transducers

Subsection ISTB of the OM Code requires that the frequency response range of vibration-measuring transducers and their readout system be from one-third of the minimum pump shaft rotational speed to at least 1,000 hertz (Hz). Licensees have proposed alternatives to this OM Code requirement in accordance with 10 CFR 50.55a(a)(3) for pumps with low shaft rotational speeds. Similar alternative requests submitted by licensees have been withdrawn following discussion with the NRC. The proposed alternatives state that the procurement and calibration of vibration-measuring transducers and their readout systems for the lower end of the OM Code-specified range were hardships because of the limited number of vendors supplying such equipment, the level of equipment sophistication, and equipment cost. The NRC typically authorized these alternative requests in the past. However, vibration-measuring transducers and their readout system can now be procured from various suppliers at a reasonably low cost due to technology advancement and research work performed in the field of instrumentation. Therefore, licensee requests to use this alternative are generally no longer authorized by the NRC.

5.14   Motor Drivers for Pumps

Pump drivers are outside the scope of the OM Code, with the exception of vibration testing for vertical line shaft pumps where the driver is an integral part of the pump. Most pumps are driven by electric motors, which are connected via coupled shafts. Motor vibration attributable to coupling misalignment may not be realized or measured at the pump, and small changes in the vibration of a motor can have significant effects on pump operation and the operational readiness.

Institute of Electrical and Electronics Engineers (IEEE) Standard 741-2007, "IEEE Standard Criteria for the Protection of Class 1E Power Systems and Equipments in Nuclear Power Generating Stations," briefly address the vibration issue, and refers to IEEE C37.96-2000, "IEEE Guide for AC Motors Protection," for motors. IEEE 741-2007 includes the testing and the surveillance requirements and lists several standards in the reference section for testing. IEEE C37.96-2000 extensively address the vibration issue on electric motors because of its significant impact on bearings, lubricants, protective devices, etc.

5.15   Pumps in New Reactors

RG 1.206 provides guidance for the development of IST programs for pumps by combined operating license (COL) licensees. In Commission papers SECY-90-016 and 93-087, the NRC indicated that new nuclear power plants should be designed to accommodate full flow testing of pumps within the scope of the IST program. As part of its review of design certification and COL applications, the NRC staff is evaluating whether new nuclear power plants will be designed to allow full flow testing of pumps in the IST program. The capability to perform full flow testing should be reflected in the development of the IST activities for pumps in new reactors.

ASME has prepared a revision to the ASME OM Code testing provisions for pumps in new reactors to provide comprehensive pump testing described in the ASME OM Code on a quarterly frequency. The updated pump testing provisions for new reactors are specified in the 2012 Edition of the ASME OM Code. Applicants for and licensees of new reactors will need to consider the ASME OM Code edition applicable to their IST program in accordance with 10 CFR 50.55a, "Codes and standards."

The staff will conduct inspections of the development and implementation of the IST program (including full flow testing of pumps) during construction and operation of new nuclear power plants.

6. STANDARD TECHNICAL SPECIFICATIONS

Standard Technical Specifications are contained in NUREG-1430, through NUREG-1434.

The Administrative Controls Technical Specification 5.5 includes a requirement to establish, implement and maintain a program entitled "Inservice Testing Program." This program provides controls for properly applying test frequencies associated with inservice testing of components activites under 10 CFR 50.55a(f) to Surveillance Requirements under 10 CFR 50.36.

7. IDENTIFICATION OF CODE NONCOMPLIANCE

7.1   Nonconforming Conditions

For details see Section 2.1.4, "Identification of Code Noncompliance," in this NUREG.

7.2 Starting Point for Technical Specification Required Action Completion Times

For details see Standard Technical Specification NUREG-1430 through NUREG-1434, "Limiting Condition for Operation (LCO) 3.0.2."

8. RISK-INFORMED INSERVICE TESTING

8.1   Introduction

10 CFR 50.55a, paragraph (f), "Inservice Testing Requirements," requires, in part, that Class 1, 2, and 3 pumps and valves must meet the requirements of the ASME OM Code.

Regulatory Guide (RG) 1.175, "An Approach for Plant-Specific, Risk-Informed Decisionmaking: Inservice Testing," describes an acceptable alternative approach for applying risk insights from probabilistic risk assessment (PRA), in conjunction with established traditional engineering information, to make changes to a nuclear power plant's IST program. The approach described in RG 1.175 addresses the high-level safety principles specified in RG 1.174, "An Approach for Using Probabilistic Risk Assessment in Risk-Informed Decisions on Plant-Specific Changes to the Licensing Basis," and attempts to strike a balance between defining an acceptable process for developing risk-informed inservice testing (IST) programs without being overly prescriptive. The resultant risk-informed IST programs will have improved effectiveness with regard to the utilization of plant resources, while still maintaining acceptable levels of quality and safety. However, licensees may propose other approaches for consideration by the NRC staff. It is intended that the approach presented in RG 1.175 should be regarded as examples of acceptable practices, and that licensees should have some degree of flexibility in satisfying regulatory requirements on the basis of their accumulated plant experience and knowledge. As discussed in RG 1.175, licensees proposing to implement a risk-informed IST program are required to submit a request to implement an alternative to the ASME OM Code in accordance with 10 CFR 50.55a(a)(3).

8.2   Discussion

Until such time as a risk-informed alternative to the current Code requirements is incorporated by reference into the regulations, the alternative approach described in RG 1.175 must be authorized by the NRC pursuant to 10 CFR 50.55a(a)(3)(i) on a plant-specific basis prior to implementation. Because 10 CFR 50.55a(a)(3)(i) places no restrictions on the scope of alternatives that may be authorized, licensees may propose risk-informed alternatives to their entire IST program, or may propose alternatives that are more limited in scope (e.g., for a particular system or group of systems, or for a particular group of components). In either case, the staff expects the licensee's proposal to address the principles described in RG 1.175, including those related to implementation and monitoring.

If a licensee proposes a risk-informed alternative to the ASME OM Code test requirements, the application should contain a summary description of the proposed alternative. The summary description should specify the key technical and administrative aspects necessary to describe and control the risk-informed alternative. The NRC staff will review and approve this summary description pursuant to 10 CFR 50.55a(a)(3)(i) and, as such, the summary description will serve as the framework within which the licensee may make future changes to its risk-informed alternative without having to resubmit it for NRC approval.

8.3   Online Inservice Testing

In an effort to shorten refueling outages, many licensees are trying to perform as much maintenance, testing, and surveillance as possible with the nuclear power plant on line.

For example, several licensees have submitted requests to obtain NRC authorization for an alternative to conduct inservice testing once per refueling cycle, rather than during the refueling outage as prescribed by the Code. In preparing (and evaluating) such alternative requests, licensees (and the NRC staff) should consider several factors to ensure that the proposed alternative provides an acceptable level of quality and safety.

If a licensee is testing a particular pump or valve during refueling outages, the licensee may have determined that it is impractical to test the pump or valve quarterly during operation. The licensee's IST program document should, therefore, discuss the basis for deferring the testing from quarterly (and during cold shutdowns) to refueling outages. Alternative requests to perform testing once each refueling cycle with the nuclear power plant on line should be prepared in light of the refueling outage justification for each affected valve or group of valves. If necessary, the licensee should revise the refueling outage justification to be consistent with the alternative request.

Licensees (and the NRC staff) should also consider whether the testing can be accomplished within the allowed outage time permitted by any applicable TS. In general, the time necessary to complete the testing should be significantly less than the allowed outage time. This is to preclude TS violations or the need to issue exigent TS amendments or notices of enforcement discretion (NOEDs). In addition, licensees should not conduct non-corrective maintenance/testing activities at power if the associated post-maintenance testing cannot reasonably be accomplished until the next outage.

Sometimes, there is a tradeoff between testing these components at power (e.g., when they could be needed to mitigate the consequences of an accident) and testing them during outages (e.g., when there may be greater reliance on shutdown cooling or when other equipment is necessarily out-of-service). Licensees should quantitatively or qualitatively address the risks associated with testing components on line, rather than testing during the refueling outage. If the proposed testing could have a significant risk impact, or if its justification includes risk-related arguments, the alternative request should be prepared and reviewed in accordance with RG 1.174, and Appendix D to SRP Chapter 19, as applicable. Licensees should also identify any compensatory measures to be established as a means to reduce the impact (e.g., risk and operational worker safety) of testing with the nuclear power plant at power.[2] If relevant, licensees should also provide information on how testing at power (rather than testing during refueling outages) will affect scheduled maintenance work windows for the applicable system (i.e., whether the testing can be completed within the work windows or whether it will extend either the shutdown or at-power work windows). In addition, licensees will need to develop a new estimate of the maintenance unavailabilities that reflects the increased maintenance activities at power, and will need to document the basis for the new estimate (e.g., use plant logs or maintenance data to include in the current estimate of the maintenance unavailabilities those activities that were being performed during shutdown that will now be performed at power).

---

[2] It should be noted that the assessment of risk resulting from performance of maintenance activities as required by 10 CFR 50.65(a)(4) of the Maintenance Rule is not sufficient justification for testing components at power. This assessment is required maintenance activities performed during power operations or during shutdowns. This configuration risk management does not address the relative merits of testing at power versus testing during refueling outage

At times, testing (or the disassembly and inspection of components) during refueling outages can be more advantageous than at-power operations from a worker safety perspective (for example, systems may be cold and depressurized). When requesting NRC authorization to perform testing with the nuclear power plant on line, licensees should consider worker safety and should discuss whether the applicable components can be adequately isolated and restored.

In Section 11.2.3 of NUMARC 93-01, Revision. 2, "Industry Guideline for Monitoring the Effectiveness of Maintenance at Nuclear Power Plants," NUMARC, now NEI provided additional guidance for conducting online maintenance and testing. It states, in part -

> Online maintenance [and testing] should be carefully managed to achieve a balance between the benefits and potential impacts on safety, reliability or availability. For example, the margin of safety could be adversely impacted if maintenance is performed on multiple equipment or systems simultaneously without proper consideration of risk, or if operators are not fully cognizant of the limitations placed on the plant due to out of service equipment. Online maintenance should be carefully evaluated, planned and executed to avoid undesirable conditions or transients, and to thereby ensure a conservative margin of core safety.

8.4   ASME Risk-Informed Code Cases

Over the past several years, ASME has developed a series of risk-informed Code Cases related to testing pumps and valves that include risk-informed provisions, including the following examples:

- OMN-1, "Alternative Rules for Preservice and Inservice Testing of Certain Motor-Operated Valve Assemblies in Light-Water Reactor Power Plants."

- OMN-3, "Requirements for Safety Significance Categorization of Components Using Risk Insights for Inservice Testing of LWR Power Plants."

- OMN-4, "Requirements for Risk Insights for Inservice Testing of Check Valves at LWR Plants."

- OMN-7, "Alternative Requirements for Pump Testing."

- OMN-11, "Motor-Operated Valve Risk-Based Inspection Code Case."

- OMN-12, "Alternative Requirements for Inservice Testing Using Risk Insights for Pneumatically and Hydraulically Operated Valve Assemblies in Light-Water Reactor Power Plants."

Certain Code Cases are listed as approved in Tables 1 and 2 of the specific revision to RG 1.192, Operation and Maintenance Code Case Acceptability: ASME OM Code," that has been incorporated by reference into 10 CFR 50.55a. Licensees may voluntarily use these Code Cases, without additional staff approval, as an alternative to complying with the ASME Code provisions that have been incorporated by reference into 10 CFR 50.55a, provided that the licensee uses the Code Cases with the conditions specified in RG 1.192 (i.e., the Code Case is generally acceptable, but the NRC staff has determined that the alternative requirements must be supplemented in order to provide an acceptable level of quality and safety).

When using ASME's risk-informed Code Cases, licensees must perform the testing and performance monitoring of individual components as specified in the risk-informed component Code Cases (e.g., OMN-1, OMN-3, OMN-4, OMN-7, OMN-11, and OMN-12), as modified by any conditions specified in RG 1.192.

The ASME OM, Subsection ISTE in the 2009 Edition to the ASME OM addresses overall aspects of risk-informed inservice testing. The NRC staff will review Subsection ISTE as part of the incorporation by reference of the 2009 Edition to the ASME OM Code in 10 CFR 50.55a with applicable conditions (if any). Note: ASME OM Code 2009 Edition details are provided for information only.

9. REFERENCES

9.1   American Society of Mechanical Engineers/American National Standards Institute NQA-1, "Quality Assurance Program Requirements for Nuclear Facilities," New York, NY, and Washington, 1979.

9.2   American Society of Mechanical Engineers (ASME) *Boiler and Pressure Vessel Code*, Section XI, "Rules for Inservice Inspection of Nuclear Power Plant Components," New York, NY.

9.3   Nuclear Energy Institute, NEI-96-07, "Guidelines for 10 CFR 50.59 Implementation," Revision 1, Washington, DC, dated November 2000, Agencywide Documents access and Management System (ADAMS) Accession No. ML003771157.

9.4   Nuclear Energy Institute White Paper, "Standard Format for Requests from Commercial Reactor Licensees Pursuant to 10 CFR 50.55a," Revision 1, Washington, DC, June 7, 2004, ADAMS Accession No. ML070100400.

9.5   The National Technology Transfer and Advancement Act of 1995, Pub. L. No. 104-113, 1995. NRC Inspection Manual Para 9900, Technical Guidance."

9.7   *U.S. Code of Federal Regulations*, "Domestic Licensing of Production and Utilization Facilities," Part 50 and Part 52, "Licenses, Certifications, and Approvals for Nuclear Power Plants," Chapter 1, Title 10, "Energy."

9.8   American Society of Mechanical Engineers/American National Standards Institute (ANSI) *Code for Operation and Maintenance of Nuclear Power Plants*, 1995 Edition through 2004 Edition including 2005 and 2006 addendas, New York, NY, and Washington, DC.

9.9   American Society of Mechanical Engineers/American National Standards Institute Standard N45.2.6, "Qualification of Inspection and Examination and Testing Personnel for Nuclear Power Plants," New York, NY, and Washington, DC.

9.10   American Society of Mechanical Engineers QME-1, "Qualification of Active Mechanical Equipments used in Nuclear Power Plants," Edition 2007, Washington, DC.

9.11   Performance Test Code, PTC 25.3-1976, "Safety and Relief Valves," New York, 1976.

9.12   American Society of Mechanical Engineers/American National Standards Institute *Operations and Maintenance Standards*, New York, NY, and Washington, DC, 1987.

9.12.1 ANSI/ASME, Part 1 (OM-1), "Requirements for Inservice Performance Testing of Nuclear Power Plant Pressure Relief Devices," New York, NY, 1981 and 1987.

9.12.2 ANSI/ASME, Part 6 (OM-6)," Inservice Testing of Pumps in Light-Water Reactor Power Plants," New York, NY, 1988 and 1989 Addenda.

9.12.3 ANSI/ASME Part 10 (OM-10), "Inservice Testing of Valves in Light-Water Reactor Power Plants," New York, NY, 1988 and 1989 Addenda.

9.13 U.S Nuclear Regulatory Commission (NRC), Regulatory Issue Summary 2004-12, "Clarification on Use of Later Editions and Addenda to the ASME OM Code and Section XI," Washington, DC, July 28, 2004.

9.14 NRC, Regulatory Issue Summary 2005-20, "Revision to NRC Inspection Manual Part 9900 Technical Guidance, 'Operability Determinations & Functionality Assessments for Resolution of Degraded or Nonconforming Conditions Adverse to Quality or Safety,'" Revision 1, Washington, DC, April 16, 2008.

9.15 NRC, Regulatory Issue Summary 2010-06, "Inservice Inspection and Testing Requirements of Dynamic Restraints (Snubbers), dated June 1, 2010.

9.16 NRC, Regulatory Issue Summary 2004-16, "Use of Later Editions and addenda to the ASME Code section XI for repair/Replacement Activities," dated October 19, 2004.

9.17 NRC Regulatory Issue Summary 2012-08, "Developing Inservice Testing and Inservice Inspection Programs under 10 CFR Part 52," dated July 16, 2012.

9.18 Nuclear Management and Resources Council (NUMARC, now NEI), NUMARC 93-01, Industry Guideline for Monitoring the Effectiveness of Maintenance at Nuclear Power Plants, Revision 2, April 1996, ADAMS Accession No. ML101020415.

9.19 NRC, Bulletin 88-04, Potential Safety-Related Pump Loss, Washington, DC, May 5, 1988.

9.20 SECY-90-016, "Evolutionary Light Water Reactor (LWR) Certification Issues and Their Relationship to Current Regulatory Requirements," January 12, 1990. ADAMS Accession No. ML003707849.

9.21 SECY-93-087, "Policy, Technical, and Licensing Issues Pertaining to Evolutionary and Advanced Light-Water Reactor (ALWR) Designs," April 2, 1993, ADAMS accession No. ML003708021.

9.22 SECY-94-084, "Policy and Technical Issues Associated with the Regulatory Treatment of Non-Safety Systems (RTNSS) in Passive Plant Designs," March 28, 1994, ADAMS Accession No. ML003708068.

9.23 SECY-95-132, "Policy and Technical Issues Associated with the Regulatory Treatment of Non-Safety Systems (RTNSS) in Passive Plant Designs (SECY-94-084),"May 22, 1995, ADAMS Accession No. ML003708005.

9.24 ASME *Boiler and Pressure Vessel Code*, New York, NY.

9.24.1 Inquiry IN 91-3, "OM Code 1987 through OMc-1990, Part 6, Para 4.6.1.1."

9.24.2 Inquiry IN 91-037, "ASME Section XI, 1977 Edition through Later Editions and Addenda through 1987 Addenda, Table IWP-4110-1, 'Instrument Accuracy Flowrate'."

9.24.3 Inquiry IN 91-045, "Section XI, IWV-3200; Valve Stroke-Time Test," March 10, 1992.

9.24.4 Inquiry IN 92-025A, "Section XI, IWV-3410 and IWV-3520; Valve Testing - Extended Shutdown," February 9, 1993.

9.24.5 Inquiry IN 92-031, "Section XI, IWA-3200; Valve Testing - Adjustment of Packing," August 27, 1992.

9.24.6 Interpretation XI-78-01.

9.24.7 Interpretation XI-1-79-19, "Section XI, Division 1, Operability Limits of Pumps, IWP-3210," File Number BC-79-150, December 12, 1979.

9.24.8 Interpretation XI-1-89-10, "Section XI, Division 1, IWV-3300, IWV-3412, and IWV-3413; Valve Exercising Test," File Number IN88-015, November 14, 1988.

9.24.9 Interpretation XI-1-89-55, "Section XI, IWP-4110, Table IWP-4110-1, and IWP-4120; Pump Instrument Accuracy," File Number IN90-021, January 9, 1991.

9.24.10 Interpretation XI-1-89-65, "Section XI, IWV-3512 and IWV-3514; Valve Set Point Adjustments," File Number IN90-002, January 15, 1991.

9.25   American Society of Mechanical Engineers/American National Standards Institute *Code for Operation and Maintenance of Nuclear Power Plants*, New York, 1990.

9.25.1 Interpretation 92-2, "OM-1-1981, Paragraphs 1.3.3.1.5, 1.3.4.1.5, and 1.3.1.3; Adjustment of Valve Setpoint - Corrective Action," File Number OMI-91-2, March 24, 1992.

9.25.2 Interpretation 92-4, "OM-1-1981, Paragraphs 8.1.1.9 and 8.1.3.8; Set Pressure Testing," File Number OMI-91-3d, March 24, 1992.

9.25.3 Interpretation 92-5, "OM-1987 With Addenda Through OMc-1990, Part 1, Paragraphs 1.1.2, 7.1.2.3, 7.2.2.3, 7.3.2.4, and 7.4.2.4; Applicability – Class 2 and 3 Vacuum Relief Valves," File Number OMI-91-4, March 24, 1992.

9.25.4 Interpretation 92-6, OM-1987 With Addenda Through OMc-1990, Part 6, Paragraph 5.2 and Table 3a: Pump Testing, File Number OMI-91-5, March 24, 1992.

9.25.5 Interpretation 93-1, OM-1987 With OMa-1988 Addenda, Part 10, Paragraphs 4.2.1 and 4.3.2; OM Code-1990, ISTC 4.2 and ISTC 4.5; Valve Testing During Extended Shutdown, File Number OMI-92-4, January 7, 1993.

9.25.6 Interpretation 95-7, OM Code-1990, ISTB 4.6.1 and Table ISTB 4.6.1.1.; OM-1987 with OMa-1988, Part 6, Para. 4.6.11 and Table 1, Instrument Accuracy.

9.26 *Federal Register*

9.26.1 *Federal Register*, Vol. 41, No. 30, "Codes and Standards for Nuclear Power Plants (10 CFR Part 50)," February 12, 1976, p. 6256.

9.26.2 *Federal Register*, Vol. 57, No. 3152, "Codes and Standards for Nuclear Power Plants (10 CFR Part 50)," August 6, 1992, p. 34666.

9.26.3 *Federal Register* Notice (56 FR 36175), July 31, 1991.

9.26.4 *Federal Register* Notice (64 FR 51370), September 22, 1999.

9.26.5 *Federal Register* Notice (66 FR 40626), August 3, 2001.

9.26.6 *Federal Register* Notice (69 FR 58804), October 1, 2004.

9.27 Generic Issues and Generic Letters

9.27.1 "Generic Evaluation of Feedwater Transients and Small Break Loss-of-Coolant Accidents in GE-Designed Operating Plants and Near-Term Operating License Applications," January 1980.

9.27.2 "Generic Issue 105, "Interfacing Systems LOCAs at Light-Water Reactors."

9.27.3 Generic Letter 87-06, "Periodic Verification of Leak-Tight Integrity of Pressure Isolation Valves," March 13, 1987.

9.27.4 Generic Letter 87-09, "Sections 3.0 and 4.0 of the Standard Technical Specifications (STS) on the Applicability of Limiting Conditions for Operation and Surveillance Requirements," May 4, 1987.

9.27.5 Generic Letter 88-14, "Instrument Air Supply System Problems Affecting Safety-Related Equipment," August 8, 1988.

9.27.6 Generic Letter 89-04, "Guidance on Developing Acceptable Inservice Testing Programs, April 3, 1989.

9.27.7 Generic Letter 89-10, Safety-Related Motor-Operated Valve Testing and Surveillance," June 28, 1989.

9.27.8 Generic Letter 90-06, "Resolution of Generic Issue 70, Power-Operated Relief Valve and Block Valve Reliability, and Generic Issue 94, Additional Low-Temperature Overpressure Protection for Light-Water Reactors, Pursuant to 10 CFR 50.54(f)," June 20, 1990.

9.27.9 Generic Letter 91-18, "Information to Licensees Regarding NRC Inspection Manual Sections on Resolution of Degraded and Nonconforming Conditions, Revision 1," October 8, 1997 (GL 91-18 superseded see RIS 2005-20).

9.27.10 Generic Letter 96-05, "Periodic Verification of Design-Basis Capability of Safety-Related Motor-Operated Valves, September 18, 1996.

9.27.11 Generic Letter 96-06, Assurance of Equipment Operability and Containment Integrity During Design-Basis Accident Conditions," September 30, 1996.

9.27.12 Generic Safety Issue 158, "Performance of Safety-Related Power-Operated Valves Under Design-Basis Conditions."

9.28 Regulatory Information Summary 2000-03, "Resolution of GSI 158 Performance of Safety-Related Valves Under Design Basis Conditions," March 15, 2000, including NRC staff comments on JOG Program on AOV periodic testing.

9.29 Information Notices and Bulletins

9.29.1 Information Notice 82-08, "Check Valve Failures on Diesel Generator Engine Cooling Systems," March 26, 1982.

9.29.2 Bulletin No. 83-03, "Check Valve Failures in Raw Water Cooling System of Diesel Generators," March 10, 1983.

9.29.3 Information Notice 83-54, "Common Mode Failure of Main Steam Isolation Non-Return Check Valves," August 11, 1983.

9.29.4 Information Notice 85-84, "Inadequate Inservice Testing of Main Steam Isolation Valves," October 30, 1985.

9.29.5 Information Notice 86-50, "Inadequate Testing to Detect Failures of Safety-Related Pneumatic Components or Systems," June 18, 1986.

9.29.6 Information Notice 87-01, "RHR [Residual Heat Removal] Valve Misalignment Causes Degradation of ECCS in PWRs," January 6, 1987.

9.29.7 Information Notice 87-40, "Back Seating Valves Routinely to Prevent Packing Leakage," August 31, 1987.

9.29.8 Information Notice 88-70, "Check Valve Inservice Testing Program Deficiencies," August 29, 1988.

9.29.9 Information Notice 89-32, "Surveillance Testing of Low-Temperature Overpressure-Protection Systems," March 23, 1989.

9.29.10 Information Notice 89-62, "Malfunction of Borg-Warner Pressure Seal Bonnet Check Valves Caused by Vertical Misalignment of Disk," August 31, 1989.

9.29.11 Information Notice 91-56, "Potential Radioactive Leakage to Tank Vented to Atmosphere," September 19, 1991.

9.29.12 Information Notice 91-74, "Changes in Pressurizer Safety Valve Setpoints Before Installation," November 25, 1991.

9.29.13 Information Notice 94-08, "Potential for Surveillance Testing To Fail To Detect an Inoperable Main Steam Isolation Valve," February 1, 1994.

9.29.14 Information Notice 94-44, "Main Steam Isolation Valve Failure To Close on Demand Because of Inadequate Maintenance and Testing," June 16, 1994.

9.29.15 Information Notice 94-56, "Inaccuracy of Safety Valve Set Pressure Determination Using Assist Devices," August 11, 1994.

9.29.16 Information Notice 96-02, "Inoperability of PORVs Masked by Downstream Indications During Testing," January 5, 1996.

9.29.17 Information Notice 96-48, "Motor-Operated Valve Performance Issues," August 21, 1996.

9.29.18 Information Notice 97-16, "Pre-Conditioning of Plant Structures, Systems, and Components before ASME Code Inservice Testing or Technical Specification Surveillance Testing," April 4, 1997.

9.29.19 Information Notice 97-90, "Use of Non-Conservative Acceptance Criteria in Safety Related Pump Surveillance Tests," December 30, 1997.

9.29.20 Information Notice 2000-21, "Detached Check Valve Disk Not Detected by the Use of Acoustic and Magnetic Non-Intrusive Test Technique," December 15, 2000.

9.29.21 Information Notice 2003-01, "Failure of a Boiling Water Reactor Main Steam Safety/Relief Valve," January 15, 2003.

9.30 RIS 2005-20, "Revision to NRC Inspection Manual Part 9900 Guidance, "Operability Determinations & Functionality Assessments for Resolution of Degraded or Nonconforming Conditions Adverse to Quality and Safety," Revision 1, dated April 16, 2008.

9.31   Inspection Manual (sections published periodically).

9.31.1 Part 9900, "Technical Guidance: Maintenance - Voluntary Entry into Limiting Conditions for Operation Action Statements To Perform Preventive Maintenance," April 18, 1991, ADAMS Accession No. 9305270187.

9.31.2 Part 9900, "Technical Guidance: Operability Determination & Functionality Assessment for Resolution of Degraded and Nonconforming Conditions Adverse to Quality or Safety," Attachment to RIS 2005-20, Revision 1, ADAMS Accession No. ML073440103.

9.31.3 Inspection Procedure 61726, "Surveillance Observations."

9.31.4 Inspection Procedure 62706, "Maintenance Rule."

9.31.5 Inspection Procedure 62708, "Motor-Operated Valve Capability."

9.31.6 Attachment 22, "Surveillance Testing," to NRC Inspection Manual IP 71111, "Reactor Safety: Initiating Events, Mitigating Systems, Barrier Integrity."

9.31.7 Inspection Procedure (IP) 73756, "Inservice Testing of Pumps and Valves," July 27, 1995,

9.31.8 Inspection Manual, Manual Chapter (MC) 0326, "Operability Determination and Functionality Assessments for Conditions Adverse to Quality and Safety, ADAMS Accession No. ML12346A480.

9.32   Letters

9.32.1 Memorandum from Frederick J. Hebdon, Director, Project Directorate II-3, Division of Reactor Projects I/II, NRC Office of Nuclear Reactor Regulation, to Jon R. Johnson, Acting Director, Division of Reactor Projects, Region II, in response to Technical Assistance Request TIA 96-007: Regulatory Acceptability of Lubricating Valves Prior to Surveillance Testing, dated July 2, 1996 (ADAMS Accession No. 9607150019)

9.32.3 Minutes of the Public Meetings on Generic Letter 89-04, October 25, 1989, ADAMS Accession No. 9001040128).

9.33   NUREGs:

9.33.1 NUREG-0737, "Clarification of TMI Action Plan Requirements," November 1980.

9.33.2 NUREG-0800, "Standard Review Plan for the Review of Safety Analysis Reports for Nuclear Power Plants."

9.33.3 NUREG-1275, "Operating Experience Feedback Report: Solenoid-Operated Valve Problems, Vol. 6," February 1991, ADAMS Accession No. 8110140308.

9.33.4 NUREG-1350, "NRC Information Digest," issued annually.

9.33.5 NUREG-1430, "Standard Technical Specifications — Babcock and Wilcox Plants,"

9.33.6 NUREG-1431, "Standard Technical Specifications — Westinghouse Plants."

9.33.7 NUREG-1432, "Standard Technical Specifications — Combustion Engineering Plants."

9.33.8 NUREG-1433, "Standard Technical Specifications — General Electric Plants (BWR/4)."

9.33.9 NUREG-1434, "Standard Technical Specifications — General Electric Plants (BWR/6)."

9.33.10 NUREG-1482, "Guidelines for Inservice Testing at Nuclear Power Plants," Revision 0 (April 1995) and Revision 1 (January 2005), ADAMS Accession Nos. 9705030476 and ML050550290.

9.34   NUREG/CPs

9.34.1 NUREG/CP-0111, "Proceedings of the Symposium on Inservice Testing of Pumps and Valves," October 1990, ADAMS Accession No. 9011060112.

9.34.2 NUREG/CP-0123, "Proceedings of the Second NRC/ASME Symposium on Valve and Pump Testing," July 1992, ADAMS Accession No. 9207270274.

9.34.3 NUREG/CP-0123, "Proceedings of the Second NRC/ASME Symposium on Valve and Pump Testing, Supplement 1," November 1992, ADAMS Accession No. 9212150075.

9.34.4 NUREG/CP-0137, "Proceedings of the Third NRC/ASME Symposium on Valve and Pump Testing," July 1994, ADAMS Accession Nos. 9408250007, and 9408030114.

9.34.5 NUREG/CP-0152, "Proceedings of the Seventh NRC/ASME Symposium on Valve and Pump Testing," July 2002, ADAMS Accession Nos. ML021970036, ML021970064, ML021970100, and ML021970132.

9.35   NUREG/CRs

9.35.1 NUREG/CR-5102, "Interfacing Systems LOCA: Pressurized-Water Reactors," February 1989, ADAMS Accession No. 8904110373.

9.35.2 NUREG/CR-5124, "Interfacing Systems LOCA: Boiling-Water Reactors," February 1989, ADAMS Accession No. 8903140208.

9.35.3 NUREG/CR-5775, "Quantitative Evaluation of Surveillance Test Intervals Including Test-Caused Risks," March 1992, ADAMS Accession No. 9203250305.

9.35.4 NUREG/CR-6644, "Generic Issue 158: Performance of Safety-Related Power-Operated Valves Under Operating Conditions," ADAMS Accession No. 9910190167.

9.35.5 NUREG/CR-5775, "Quantitative Evaluation of Surveillance Test Intervals Including Test-Caused Risks," Dated February 1992, ADAMS Accession No. ML027410457.

9.35.6 NUREG/CR-6396, "Examples, Clarifications, and Guidance on Preparing Request for Relief from Pump and Valve Inservice Testing Requirements," Prepared by Idaho National Engineering Laboratory and published on February 1996, ADAMS Accession No. 9603010300.

9.36   Regulatory Guides (RGs)

9.36.1 Regulatory Guide 1.26, "Quality Group Classifications and Standards for Water-, Steam-, and Radioactive-Waste-Containing Components of Nuclear Power Plants," Revision 4, March, 2007.

9.36.2 Regulatory Guide 1.58, "Qualification of Nuclear Power Plant Inspection, Examination, and Testing Personnel," Revision 1, September 1980. (Withdrawn - See 56 FR 36175, July 31, 1991)

9.36.3 Regulatory Guide 1.147, "Inservice Inspection Code Case Acceptability ASME Section XI, Division 1," Revision 16, October 2010.

9.36.4 Regulatory Guide 1.174, "An Approach for using Probabilistic Risk Assessment in Risk-Informed Decisions on Plant-Specific Changes to the Licensing Basis," Revision 2, May 2011.

9.36.5 Regulatory Guide 1.175, "An Approach for Plant-Specific, Risk-Informed Decision-Making: Inservice Testing," August 1998.

9.36.6 Regulatory Guide 1.192, "Operation and Maintenance Code Case Acceptability: ASME OM Code," June 2003.

9.36.7 Regulatory Guide 1.193, Revision 3, "ASME Code Cases Not Approved for Use," October, 2010.

9.36.8 Regulatory Guide 1.187, "Guidance for Implementation of 10 CFR 50.59, Changes, Tests, and Experiments," Washington, DC, November 2000.

9.36.9 Regulatory Guide 1.206, "Combined Licensee Application for Nuclear Power Plants (LWR Edition)," June 2007.

9.36.10 Regulatory Guide 1.100, "Seismic Qualification of Electric and Mechanical Equipments for Nuclear Power Plants," September 2009.

9.37   Regulatory Issue Summary 2004-12, "Clarification on use of Later Editions and Addenda to the ASME OM Code and Section XI."

9.38   Regulatory Issue Summary 2012-08, "Developing Inservice Testing and Inservice Inspection Programs Under 10 CFR Part 52."

9.39   SECY-77-439, "Single-Failure Criterion," August 17, 1977, ADAMS Accession No. 7812180291.

9.40   SECY-92-223, "Resolution of Deviations Identified during the Systematic Evaluation Program," June 19, 1992, ADAMS Accession No. 9206300320.

9.41   Supplement to Minutes of the Public Meetings on Generic Letter 89-04, September 16, 1991, ADAMS Accession No. 9001040128.

9.42   WASH-1400 (NUREG-75/014), "Reactor Safety Study: An Assessment of Accident Risk in U.S. Commercial Nuclear Power Plants," 1975 (ADAMs Accession No. ML053290245).

APPENDIX A

Guidelines for Inservice Examination and Testing Program for Dynamic Restraints (Snubbers) at Nuclear Power Plants

CONTENTS

APPENDIX A

Figures

Tables

ABBREVIATIONS

ASME	American Society of Mechanical Engineers
B&PV	Boiler and Pressure Vessel
BWR	boiling-water reactor
CFR	Code of Federal Regulations
DBD	design-basis document
DTPG	defined test plan group
EGM	enforcement quidance memorandum
FR	Federal Register
FSAR	final safety analysis report
GDC	General Design Criterion
GL	generic letter
GSI	generic safety issue
IN	information notice
IP	inspection procedure
ISI	inservice inspection
IST	inservice testing
ITAAC	inspection, tests, analyses, and acceptance criteria
LBHS	large bore hydraulic snubbers
LCO	limiting condition operation
NEI	Nuclear Energy Institute (formerly NUMARC)
NRC	U.S. Nuclear Regulatory Commission
NRR	Office of Nuclear Reactor Regulation (NRC)
NUMARC	Nuclear Management and Resources Council (now NEI)
OM	Operations and Maintenance
P&ID	piping and instrument diagram
PWR	Pressurized-water reactor
RCP	reactor coolant pump
RG	regulatory guide
RIS	Regulatory Issue Summary
SAR	safety analysis report
SG	steam generator
SLM	service-life monitoring
STS	Standard Technical Specifications

TS	Technical Specification(s)
TRM	Technical Requirement Manual
UFSAR	updated final safety analysis report
WGC	Working Group Committee

1. INTRODUCTION

1.1 Regulatory Basis

Title 10, Section 50.55a, of the Code of Federal Regulations (10 CFR 50.55a), "Codes and Standards," defines the requirements for applying industry codes and standards to boiling- or pressurized-water-cooled nuclear power facilities. Each of these facilities is subject to the conditions in paragraphs (a), (f), and (g) of 10 CFR 50.55a, as they relate to inservice inspection and inservice testing. By rulemaking effective September 8, 1992 (*see* 57 FR 34666: August 6, 1992), the U.S. Nuclear Regulatory Commission (NRC) established paragraph (f) of 10 CFR 50.55a to separate the inservice testing (IST) requirements from the inservice inspection (ISI) requirements in paragraph (g). The inservice testing requirements for pumps and valves fall under paragraph (f), whereas inservice examination and testing requirements for snubbers fall under paragraph (g).

As of July 11, 2011, the NRC regulations in 10 CFR 50.55a(b)(2) incorporate by reference the 1970 Edition through the 2007 Edition with 2008 addenda of the American Society of Mechanical Engineers (ASME) *Boiler and Pressure Vessel* Code (B&PV), Section XI, "Rules for Inservice Inspection of Nuclear Power Plant Components," in which Article IWF-5000 specifies the requirements of inservice examination and testing of snubbers. Similarly as of July 11, 2011, the NRC regulations in 10 CFR 50.55a(b)(3) incorporate by reference the 1995 Edition through the 2004 Edition with 2005 and 2006 addendas of the *Code for Operation and Maintenance of Nuclear Power Plants* (OM Code) promulgated by the ASME, in which Subsection ISTA of the OM Code provides general requirements for IST of pumps and valves and inservice examination and testing of snubbers. Subsections ISTB, ISTC, and ISTD specify the inservice requirements for pumps, valves, and dynamic restraints, respectively. Based on those requirements, each of the nuclear power plant licensees must establish an IST program for pumps and valves and an inservice examination and testing program for snubbers, specify the components included in the program as well as the applicable examination and test methods and frequencies for those components, and implement the program in accordance with the OM Code.

Where a snubber examination and test requirements of the ASME B&PV or ASME OM Code is determined to be impractical for a facility, the NRC's regulations allow the licensee to submit a request for relief from the given requirement, along with information to support the determination. Relief requests generally detail the reasons for deviating from the Code requirements and propose alternative testing methods or frequencies. The Commission is authorized to evaluate licensees' relief requests, and may grant the requested relief or impose alternative requirements, considering the burden that the licensee might incur if the Code requirements were enforced for the given facility. Pursuant to 10 CFR 50.55a(a)(3)(i) and (ii), the Commission may also authorize the licensee to implement an alternative to the Code requirements, provided that the alternative ensures an acceptable level of quality and safety or the Code requirement presents a hardship without a compensating increase in the level of quality and safety.

The regulations in 10 CFR 50.55a(g)(4)(iv) specify that inservice examination of components and system testing of components (which includes snubbers) may meet the requirements in editions and addenda of the ASME B&PV Code Section XI or OM Code that were published more recently than those that are incorporated by reference in 10 CFR 50.55a(b), subject to Commission approval and the limitations and modifications listed in 10 CFR 50.55a(b). Requests for approval to use later editions and addenda previously incorporated by reference in 10 CFR 50.55a may be made via letter to the NRC. For further clarification, see NRC Regulatory Issue Summary (RIS) 2004-12, "Clarification on Use of Later Editions and Addenda to the ASME OM Code and ASME B&PV Code Section XI."

The regulation in 10 CFR 50.55a(g) establishes the inservice inspection (ISI) requirements, including the effective edition and addenda of the ASME B&PV Code that licensees must use when performing ISI of components (including supports). 10 CFR 50.55a(g)(4) states, "Throughout the service life of a boiling or pressurized water-cooled nuclear power facility, components (including supports) which are classified as ASME Code Class 1, Class 2, and Class 3 must meet the requirements, except design and access provisions and preservice examination requirements, set forth in ASME B&PV Code Section XI of editions of the ASME B&PV Code and addenda." ASME B&PV Code, Section XI provides the rules for ISI of nuclear power plant components.

The regulations in 10 CFR 50.55a(g)(4)(ii) require the use of the latest edition of the Code and addenda that has been incorporated by reference 12 months prior to the beginning of each 120-month inspection interval. This Code is considered to be the "Code of Record" for the inspection interval.

10 CFR 50.55a(g)(4)(iv) states that ISI of components (including supports) may meet the requirements set forth in subsequent editions to the "Code of Record" and addenda that are incorporated by reference in 10 CFR 50.55a(b), subject to limitations and modifications listed in 10 CFR 50.55a(b) and subject to Commission approval.

The ISI of ASME Code Class 1, 2, and 3 components including snubbers shall be performed in accordance with Section XI of the ASME B&PV Code and applicable addenda as required by 10 CFR 50.55a(g), except where specific written relief has been granted by the NRC, pursuant to 10 CFR 50.55a(g)(6)(i). Section 50.55a(a)(3) states that alternatives to the requirements of paragraph (g) may be used, when authorized by the NRC, if: (i) the proposed alternatives would provide an acceptable level of quality and safety, or (ii) compliance with the specified requirements would result in hardship or unusual difficulty without a compensating increase in the level of quality and safety.

10 CFR 50.55a(b)(3)(v) allows the optional use of Subsection ISTD, "Preservice and Inservice Examination and Testing of Snubbers in Light-Water Reactor Nuclear Power Plants," of the ASME OM Code-1995 Edition through the latest edition and addenda incorporated by reference in 10 CFR 50.55a(b)(3) in lieu of ASME B&PV Code, Section XI, Articles IWF-5200(a) and (b) and IWF-5300(a) and (b) by making appropriate changes to technical specification (TS) or licensee-controlled documents. The regulation at 10 CFR 50.55a(b)(3)(v) also states, "Preservice and inservice examination must be performed using the VT-3 visual examination method described in IWA-2213." Licensees using ASME OM Code -1995 through 2004 Edition should use IWA-2213 of the ASME B&PV Code, Section XI, 2004 Edition. Licensees using later

editions of the ASME OM Code should use IWA-2213(a) of the applicable later edition of the ASME B&PV Code, Section XI to which the licensee is committed. Note: The 2003 Addenda of ASME B&PV Code, Section XI as published included an unapproved change to IWA-2213 (which was later incorporated in post-2004 editions). If a licensee is using ASME B&PV Code Section XI, 2003 addenda as its "Code of Record," the licensee may use IWA-2213(a) (as originally published in 2003 Addenda of ASME B&PV Code, Section XI), because ASME B&PV Code, Section XI issued an Errata to correct this error via Volume 53 of ASME B&PV Code, Section XI Interpretations.

1.2 Regulatory History of Staff Guidance on Examination and Testing of Snubbers

Since the start of commercial operation of the nuclear power plants, inservice visual inspection and functional testing of snubbers have been regulatory requirements. Originally, these requirements were imposed by the plant TS surveillance requirements (SRs). There are two inservice SRs requirements for snubbers (1) visual inspection or examination; and (2) functional or operational testing. The visual inspection is the observation of the condition of installed snubbers to identify those that are damaged, degraded, or inoperable as caused by physical means, leakage, corrosion, or environmental exposure. TS surveillance testing utilizes statistical sampling to validate the functionality of the tested population within an assumed quality confidence level. Only a sample of the installed snubbers is actually tested. Functional testing is to verify that the tested snubber can operate within the specified performance limits. The snubber functional test is performed to check its operational-readiness. The functional test typically involves removing the snubber from service and testing it on a specially-designed test stand. The performance of visual examinations is a separate process that complements the functional testing program and provides additional confidence in snubber population reliability. The TS specifies a schedule for snubber visual inspections and functional testing, which is usually based on refueling outage intervals.

Improved TS for various boiling and pressurized water-cooled nuclear power plants (NUREG-1430 thru 1434, Revision 3) allow relocating inservice examination and testing requirements of snubbers from the TS to a plant's Technical Requirements Manual (TRM). Relocating snubber ISI and testing requirements from the TS to TRM however, does not eliminate the need to comply with the 10 CFR 50.55a requirements.

In 1990, the NRC issued Generic Letter (GL) 90-09, "Alternate Requirements for Snubber Visual Inspection Intervals and Corrective Actions." The alternative visual inspection schedule is based on the number of unacceptable snubbers found during the previous inspection interval in proportion to the sizes of the various snubber populations or categories. This reduces future occupational radiation exposure and is highly cost effective. The alternative inspection interval is based on a fuel cycle of up to 24 months.

At that time most of the licensees revised their snubber examination and testing documents such as TS, TRM or licensee-controlled documents, based on GL 90-09. GL 90-09 provides an alternative to requirements for snubber visual inspection intervals only.

Until 1990, the ASME Code requirements addressing inservice examination and testing of snubbers were only contained in ASME B&PV Code, Section XI, Article IWF-5000. IWF-5000 referenced

ASME/American National Standards Institute Standard for Operation and Maintenance of Nuclear Power Plants, Part 4 (OM-4), 1987 Edition with OMa-1988 Addenda for the preservice and inservice examinations and testing of snubbers. In 1990, the ASME published the initial edition of the ASME OM Code which provides rules for inservice examination and testing of snubbers by incorporating most of the requirements of OM-4. The 1995 Edition of the ASME OM Code incorporated all the visual inspection requirements provided in GL 90-09. Examination programs in accordance with ASME B&PV Code Section XI, Article IWF-5000 are required to meet the requirements of OM-4 (1987/88 addenda), which does not incorporate the alternative GL 90-09 intervals. Licensees wishing to utilize the GL 90-09 alternative intervals for visual examinations while under the IWF-5000 (OM-4) requirements must request relief from the OM-4 interval requirements.

The 1990 Edition of the ASME OM Code consisted of one section (Section IST) entitled "Rules for Inservice Testing of Light-Water Reactor Power Plants." This section is divided into four subsections: ISTA, "General Requirements," ISTB, "Inservice Testing of Pumps in Light-Water Reactor Power Plants," ISTC, "Inservice Testing of Valves in Light-Water Reactor Power Plants," and ISTD, "Examination and Performance Testing of Nuclear Power Plant Dynamic Restraints (Snubbers)." At that time, the inservice examination and testing of snubbers was governed (under rulemaking) by the ISI requirements of Section XI of the ASME B&PV Code-1989 Edition. Therefore, Subsection ISTD was not incorporated by reference in 10 CFR 50.55a at that time.

In 2000, for the first time, the proposed rule 10 CFR 50.55a(b)(3)(v) stated that licensees may use the guidance in Subsection ISTD, ASME OM Code, 1995 Edition with the 1996 Addenda, for examination and testing of snubbers. The current regulation at 10 CFR 50.55a(b)(3)(v) allows the optional use of Subsection ISTD, "Preservice and Inservice Examination and Testing of Dynamic Restraint (Snubbers) in Light-Water Reactor Nuclear Power Plants," of the ASME OM Code-1995 Edition through the latest edition and addenda incorporated by reference in 10 CFR 50.55a(b)(3) in lieu of ASME B&PV Code, Section XI, Articles IWF-5200(a) and (b) and IWF-5300(a) and (b) by making appropriate changes to TS or licensee-controlled documents. The regulation at 10 CFR 50.55a(b)(3)(v) also states, "Preservice and inservice examination must be performed using the VT-3 visual examination method described in IWA-2213." The licensees using ASME OM Code -1995 through 2004 Edition should use IWA-2213 of the ASME B&PV Code Section XI, 2004 Edition. Licensees using a later edition of the ASME OM Code should use IWA-2213(a) of the applicable later edition of ASME B&PV Code Section XI. Note: The 2003 Addenda of ASME Section XI as published included an unapproved change to IWA-2213 (which was later incorporated in post-2004 editions). If the licensee is using ASME B&PV Code Section XI, 2003 addenda as its "Code of Record," the licensee may use IWA-2213(a) (as originally published in the 2003 Addenda of ASME B&PV Code Section XI), because ASME B&PV Code Section XI issued an Errata to correct this error via Volume 53 of ASME Section XI Interpretations.

Some commenters proposed Subsection ISTD as an acceptable alternative to all preservice and inservice examination requirements in IWF-5000, ASME B&PV Code Section XI. The NRC has not accepted this suggestion because some preservice and inservice examinations for snubbers are not included in the OM Code. For example, Subsection ISTD does not address the scope of IWF-5200(c) and IWF-5300(c), inspection of integral and non-integral attachments, such as lugs, bolting, pins, and clamps. ISTA governs the scope for the ISTD and it does address the integrity of reactor coolant pressure boundary, which was added in ASME OMa-2002, addenda to the OM Code-2002.

The guidance in this NUREG revision is applicable to the 2004 Edition with 2005 and 2006 addendas of the ASME OM Code and 2007 Edition with 2008 addenda of the ASME B&PV Code, Section XI.

Snubber inservice inspection provisions are specified in the editions and addenda of ASME B&PV Code Section XI up through the 2005 Addenda. Snubber inservice inspection provisions were removed from ASME B&PV Code Section XI in the 2006 Addenda. Snubber inservice inspection provisions are also located in Subsection ISTD of the ASME OM Code, and 10 CFR 50.55a(b)(3)(v) allows licensees the option of using the inservice examination and testing provisions for snubbers in ASME B&PV Code Section XI or the ASME OM Code. However, the ASME B&PV Code Section XI option will no longer exist when using the 2006 addenda and later editions and addenda of ASME B&PV Code Section XI because these editions and addenda of Section XI do not provide inservice inspection provisions for snubbers. When using the 2006 addenda or later editions of ASME B&PV Code Section XI, snubber examination and testing must be in accordance with the ASME OM Code, Subsections ISTA and ISTD, or relief must be obtained from NRC.

1.3   NRC Recommendations and Guidance

The recommendations herein supplement the guidance provided in the ASME B&PV Code, Section XI and ASME OM Code for inservice inspection and testing of snubbers. This document is written based on the requirements as specified in the 2004 Edition with 2005 and 2006 addendas of the ASME OM Code and 2007 Edition with 2008 addenda of the ASME B&PV Code, Section XI, which are the latest edition of the ASME OM Code and ASME B&PV Code, Section XI incorporated into Paragraph (b) of 10 CFR 50.55a. To the extent practical, this document reflects the applicable section, subsection, or paragraph of the applicable documents (subsection of 10 CFR 50.55a, OM Code, ASME B&PV Code Section XI, regulatory guides, etc.).

The guidance presented herein may be applied when requesting to use relief and alternatives in lieu of the ASME Code requirements. However, licenses may also request relief that is not in conformance with the guidance. The NRC may reference a recommendation contained in this document in future safety evaluations and may grant relief or authorize an alternative if the licensee has addressed all aspects included in the respective section, where applicable.

This document specifically discusses applicable portions of Article IWA-1000 and IWA-2000, IWA-4000, IWF-5000, and IWF-6000 of the ASME B&PV Code, Section XI and Subsections ISTA and ISTD, Nonmandatory Appendices A (with Supplement 3) through H of the ASME OM Code, which licensees may implement pursuant to 10 CFR 50.55a(g)(4)(iv). It also provides guidance for licensees to use 10 CFR 50.55a(g)(4)(iv) in updating their snubber ISI and testing program to the requirements of the ASME B&PV Code, Section XI or OM Code as applicable.

If the licensee chooses to implement the guidance contained herein for issues approved under 10 CFR 50.55a(g)(4)(iv), any deviation from the guidance requires Commission approval.

1.4   Synopsis of Report

This appendix follows the format of a typical ISI and testing snubber program plan, including Development and Implementation, General Guidance, and Code Noncompliance.

Section 2, "Developing and Implementing an Inservice Examination and Testing Program of Snubbers," describes existing inservice examination and testing requirements, discusses the scope of the snubber program, and describes guidance for presenting information in snubber programs. This also provides specific recommendations on snubber and large bore (steam generator and reactor coolant pump) snubber related issues.

Section 3 and 4 present a list of references related to snubbers.

These discussions are intended to clarify the existing requirements of the Code or the regulations and, as such, they may provide recommendations to ensure that licensees continue to meet the Code and other regulatory requirements.

2. DEVELOPING AND IMPLEMENTING AN INSERVICE EXAMINATION AND TESTING PROGRAM OF SNUBBERS

Licensees may use the following guidance for developing and implementing snubber inservice examination and testing programs. This guidance supplements existing requirements and previously approved guidance on inservice examination and testing.

2.1 Compliance Considerations

The NRC regulations in Section 50.55a state that ISI of components (including supports) which are classified as ASME Code Class 1, Class 2, and Class 3 must meet the requirements, except design and access provisions and preservice examination requirements, set forth in Section XI of editions of ASME B&PV Code and addenda. ASME B&PV Code, Section XI provides the rules for ISI of nuclear power plant components. 10 CFR 50.55a requires ISI of components (including supports) without specifically mentioning "snubbers." 10 CFR 50.55a also allows the optional use of Subsection ISTD of the OM Code promulgated by the ASME for inservice inspection and testing of snubbers. This ISI along with testing is intended to assess the reliability and operational readiness of the snubbers.

10 CFR 50.55a(g)(4)(ii) requires the use of the latest edition and addenda of the Code that has been incorporated by reference 12 months prior to the beginning of each 120-month inservice interval. This Code is considered to be the "Code of Record" for the inspection interval. 10 CFR 50.55a(g)(4)(iv) allows inservice inspection of components (including supports) to meet the requirements set forth in subsequent editions to the "Code of Record" and addenda that are incorporated by reference in 10 CFR 50.55a(b), subject to limitations and modifications listed in 10 CFR 50.55a(b) and subject to NRC approval. Request for approval to use later editions and addenda previously incorporated by reference in 10 CFR 50.55a may be made via a letter to the NRC. See RIS 2004-12 for further clarification. However, pursuant to 10 CFR 50.55a(g)(4)(iv), licensees' ISI and testing programs may meet the requirements of editions and addenda of the Code (or portions thereof) that are more recent than those incorporated in 10 CFR 50.55(b). When requesting to use editions and addenda of the ASME Code that have not been incorporated by reference, licensees must request authorization to use these later editions and addenda as an alternative to the regulations pursuant to 10 CFR 50.55a(a)(3). When licensees choose to use any or all portions of a revised edition, they must meet all related requirements of the respective editions or addenda, and such exceptions are subject to NRC approval in accordance with 10 CFR 50.55a(g)(4)(iv).

The NRC may authorize alternatives to Code testing requirements submitted as relief requests or in a similar format that includes a discussion of the requirements, a description of the proposed alternative, and the justification for approval of the alternative. 10 CFR 50.55a includes the following provisions for authorizing alternatives or granting relief:

- 10 CFR 50.55a(a)(3)(i) allows the NRC to authorize alternatives if the proposed alternatives would provide an acceptable level of quality and safety. The NRC will normally authorize an alternative pursuant to this provision only if the licensee proposes a method of testing that is equivalent to, or an improvement of, the method specified by the Code, or if the testing will comply or is consistent with later Code editions approved by the NRC in 10 CFR 50.55a(b).

- 10 CFR 50.55a(a)(3)(ii) allows the NRC to authorize an alternative if compliance with the Code requirement would result in hardship or unusual difficulty without a compensating increase in the level of quality and safety. The NRC may authorize an alternative pursuant to this provision if, although the proposed alternative testing does not comply with the Code, the increase in overall plant safety and quality attained by complying with the Code requirement is not justified in light of the difficulty associated with compliance.

- 10 CFR 50.55a(f)(6)(i) includes the following provision:
  The Commission will evaluate determinations that Code requirements are impractical. The Commission may grant relief and may impose such alternative requirements as it determines is authorized by law, giving due consideration to the burden upon the licensee that could result if the requirements were imposed on the facility.

The NRC may grant relief pursuant to this provision or may impose alternatives if the licensee demonstrates that the design or access limitations make the Code requirement impractical. Thus, the staff's evaluation considers the burden created by imposing the Code requirements on the licensee.

For plants using their TS to govern ISI and testing of snubbers, 10 CFR 50.55a(g)(5)(ii) requires that if a revised ISI program for a facility conflicts with the TS, the licensee shall apply to the NRC for amendment of the TS to conform the TS to the revised program. Therefore, when performing 120-month ISI program updates in accordance with 10 CFR 50.55a (g)(4), licensees must submit any required amendments to ensure their TS remain consistent with the new Code of record or NRC-approved alternative used in lieu of the Code requirements. The TS governing the snubber ISI and test program do not eliminate the 10 CFR50.55a requirements to update the program at 120-month intervals or to request and receive NRC authorization for alternatives to the Code requirements when appropriate.

2.1.1   ASME Code Case Applicability

A Code Case is the official method of the ASME for handling a reply to an inquiry when study indicates that the Code wording needs clarification, or when the reply modifies the existing requirements of the Code, or grants permission to use alternative methods. ASME develops Code Cases through a consensus process to clarify the intent of existing Code requirements or to provide an alternative to a specific Code requirement. A Code Case may be issued for the purpose of providing alternative rules when justified, to permit early implementation of an approved revision when the need is urgent, or to provide rules not covered by existing provisions of the ASME OM Code.

The NRC reviews new or revised Code Cases to determine their acceptability for incorporation by reference in 10 CFR 50.55a through the subject regulatory guides. Accordingly, the NRC staff developed RG 1.192, "Operation and Maintenance Code Case Acceptability, ASME OM Code," as well as RG 1.193, "ASME Code Cases Not Approved for Use."

The regulations at 10 CFR 50.55a(b)(6) incorporate by reference RG 1.192. Licensees may implement the code cases listed in RG 1.192 without obtaining further NRC review or approval if the Code Cases are used in their entirety with any supplemental conditions specified in the RG and the licensee's IST Code of Record is applicable to the Code Case. RG 1.193 lists Code Cases not approved for use.

If a licensee would like to use an ASME Code Case with an Edition or Addendum of the ASME Code to which it is not applicable, the licensee has the following options:

a.   Have the alternative to use the Code Case, beyond its stated applicability, authorized by the NRC pursuant to 10 CFR 50.55a(a)(3), or

b.   If the Code Case is applicable to an Edition or Addendum of the ASME Code later than the version of the Code being used by the licensee, the licensee could update to the later version of the Code pursuant to 10 CFR 50.55a(f)(4)(iv) or (g)(4)(iv) and then use the Code Case, provided the Code Case has been approved for use in the appropriate Regulatory Guide and incorporated by reference into 10 CFR 50.55a. Note that the later version of the ASME Code must also have been incorporated by reference into 10 CFR 50.55a, the licensee must update all related requirements of the respective Edition or Addenda, and the update must be specifically approved by the Commission.

The NRC may authorize the use of a Code Case that has not yet been approved for use in RG 1.192 if a licensee requests the use of the code case under 10 CFR 50.55a(a)(3). The NRC may authorize the use of such a Code Case until a future revision to RG 1.192 accepts the use of the ASME code case. At that time, if the licensee intends to continue implementing the Code Case, they must follow all the provisions of the Code Case with the conditions specified in RG 1.192, if any. The authorization for a specific licensee to use a Code Case that is not listed in RG 1.192 does not authorize any other licensee to use the Code Case without submittal by the subsequent licensee of a request to implement an alternative to the ASME OM Code requirements under 10 CFR 50.55a(a)(3).

If RG 1.193 identifies a Code Case as being unacceptable, the NRC is unlikely to approve a licensee request to use that specified Code Case (whether by exemption, approval of alternatives, or authorizing relief). Licensees requesting the NRC's approval to implement a Code case listed in the RG 1.193 must show, at minimum, that adequate protection to public health and safety is provided if the Code Case is applied by the licensee/applicant.

An inservice examination and testing snubber program, including implementing procedures, is subject to the requirements of 10 CFR Part 50, Appendix B, "Quality Assurance Criteria for Nuclear Power Plants and Fuel Reprocessing Plants," and ASME OM Code, Subsection ISTA or ASME B&PV Code, Section XI, Section IWA. Changes to the scope, test methods, or acceptance criteria shall be reviewed to the requirements of 10 CFR 50.59, "Changes, Tests, and Experiments," 10 CFR 50.55a, and 10 CFR 50.65, "Requirements for Monitoring the Effectiveness of Maintenance at Nuclear Power Plants," as appropriate.

2.1.2 Conditions to the ASME OM Code

2.1.2.1 <u>10 CFR 50.55a(b)(3)(v)—Subsection ISTD</u>

This condition provides requirements for the examination and testing of snubbers. The condition in 10 CFR 50.55a(b)(3)(v) allows licensees using editions and addenda up to the 2005 Addendum of the ASME B&PV Code Section XI, to optionally use Subsection ISTD of the OM Code in place of the requirements for snubbers in Section XI. This condition also states that snubber preservice and inservice examinations must be performed using the VT-3 visual examination method (as described in IWA-2213 or IWA-2213(a)) when using Subsection ISTD of the OM Code. The NRC imposed the VT-3 visual examination requirement to ensure that licensees use an appropriate visual examination method for the inspection of integral and nonintegral snubber attachments, such as lugs, bolting, and clamps, when using Subsection ISTD.

Licensees that use the 2006 Addendum and later editions and addenda to Section XI of the ASME B&PV Code must follow the requirements of Subsection ISTD of the OM Code for snubbers because snubber inservice examination and testing requirements have been deleted from the scope of ASME B&PV Code, Section XI in the 2006 Addendum. The condition in 10 CFR 50.55a(b)(3)(v) does not invoke the VT-3 visual examination requirement when licensees use the 2006 Addendum and later editions and addenda to Section XI because ASME revised Figure IWF-1300-1 in the 2006 Addendum to Section XI to clarify that integral and nonintegral snubber attachments are within the scope of Section XI. Therefore, the visual examination method specified in the 2006 Addendum and in later editions and addenda to ASME B&PV Code, Section XI applies to the examination of integral and nonintegral snubber attachments.

2.1.3 Program Controls

Some licensees have incorrectly interpreted that the examination and testing of snubbers is not a 10 CFR 50.55a requirement because (1) 10 CFR 50.55a(g) addresses components (including supports) without mentioning snubbers, (2) snubber examination and testing was historically covered by TS, and (3) TS allow snubber examination and testing requirements to be relocated from the TS to the TRM. Licensees have the option to control the inservice examination and testing of snubbers through their TS or other licensee-controlled documents. For plants using their TS to govern the inservice examination and testing of snubbers, 10 CFR 50.55a(g)(5)(ii) requires that if a revised ISI program for a facility conflicts with the TS, the licensee shall apply to the Commission for the amendment of the TS to conform the TS to the revised program. Therefore, when performing 120-month program updates in accordance with 10 CFR 50.55a(g)(4), licensees must submit any required amendments or any alternative requests to ensure that their TS remain consistent with the new ISI program. The TS, TRM, or other licensee-controlled documents governing the snubber inservice examination and testing program do not eliminate the 10 CFR 50.55a requirements to update the program at 120-month intervals in accordance with 10 CFR 50.55a(g)(4) or to request and receive NRC authorization for alternatives to the Code requirements when appropriate. The NRC issued Enforcement Guidance Memorandum (EGM) 2010-01, and Regulatory Issue Summary (RIS) 2010-06, "Inservice Inspection and Testing Requirements of Dynamic Restraints (Snubbers)," on June 1, 2010, to inform the licensees, and clarify NRC's rules and regulations regarding snubber inservice examination and testing, in accordance with 10 CFR 50.55a, at nuclear power facilities.

2.1.4   OM Part 4 Clarification

The NRC has noted that the relocation of the reference to ASME/ANSI *Operation and Maintenance of Nuclear Power Plants* (OM Part 4) from IWF-5000 of ASME B&PV Code, Section XI to Table IWF-1600-1 of Section XI has created confusion regarding which edition and addenda of OM Part 4 must be used. For clarification, the ASME OM Code and OM Part 4 are two different ASME documents. Article IWF-5000 of the 1987 Addendum through the 1992 Edition of Section XI requires that the inservice examination and testing of snubbers be accomplished in accordance with the 1987 Edition of OM Part 4. The reference to OM Part 4 was deleted from IWF-5000 in the 1992 Addendum of Section XI and there is no reference to OM Part 4 in IWF-5000 in the 1992 through 2005 Addenda of Section XI. The reference for the applicable edition and addenda of OM Part 4 was moved to Table 1600-1 in the 1992 Addendum of Section XI. Although IWF-5000 in the 1992 through 2005 Addenda of Section XI no longer references OM Part 4, Table 1600-1 of the 1992 through 2005 Addenda of Section XI requires that inservice examination and testing of snubbers be performed in accordance with the 1987 Edition and 1988 Addendum of OM Part 4.

Snubber inservice inspection provisions are specified in the editions and addenda of Section XI up through the 2005 Addenda. Snubber inservice inspection provisions were removed from Section XI in the 2006 Addendum. Snubber inservice inspection provisions are also located in Subsection ISTD of the ASME OM Code, and 10 CFR 50.55a(b)(3)(v) allows licensees the option of using the inservice inspection provisions for snubbers in Section XI or the ASME OM Code. However, ASME B&PV Code, Section XI option will no longer exist when using the 2006 addendum and later editions and addenda of Section XI because these editions and addenda of Section XI do not provide inservice inspection provisions for snubbers. When using the 2006 addendum or later editions of ASME B&PV Code, Section XI, snubber examination and testing must be in accordance with the ASME OM Code, Subsections ISTA and ISTD or relief must be obtained from the NRC.

2.2   Scope of Inservice Examination and Testing Programs

General Design Criterion (GDC) 1, "Quality Standards and Records," of Appendix A, "General Design Criteria for Nuclear Plants," to the 10 CFR Part 50 requires that all structures, systems, and components that are necessary for safe operation must be tested to demonstrate that they will perform satisfactorily in service. Among other things, GDC 1 requires that components that are important to safety must be tested to quality standards that are commensurate with the importance of the safety function(s) to be performed. Appendix B to 10 CFR Part 50 describes the requisite quality assurance program, which includes testing, for safety-related components. In addition, 10 CFR 50.55a(g) requires that licensees must use the ASME B&PV Code, Section XI or the optional ASME OM Code for inservice examination and testing of components that are covered by the Code. Each licensee has the responsibility to demonstrate the continued operability of all components within the scope of their snubber inservice examination and testing program. The regulatory guides augment those requirements by providing additional NRC guidance regarding scope and classification. In short, the ASME Code defines the scope,

10 CFR 50.55a endorses the Code with clarifications, and regulatory guides provide additional guidance.

2.2.1 Basis for Scope Requirements

The regulations at 10 CFR 50.55a(g) establish the ISI requirements that licensees must satisfy when performing ISI of components (including supports). Specifically, 10 CFR 50.55a(g)(4) states, "Throughout the service life of a boiling or pressurized water-cooled nuclear power facility, components (including supports) which are classified as ASME Code Class 1, Class 2, and Class 3 must meet the requirements, except design and access provisions and preservice examination requirements, set forth in Section XI of editions of the ASME B&PV Code and addenda." ASME B&PV Code, Section XI provides the rules for ISI of nuclear power plant components.

The regulation at 10 CFR 50.55a(g)(4)(ii) requires the use of the latest edition and addenda of the Code that has been incorporated by reference 12 months prior to the beginning of each 120 month inspection interval. This Code is considered the "Code of Record" for the inspection interval.

The regulation at 10 CFR 50.55a(g)(4)(iv) states that ISI of components (including supports) may meet the requirements set forth in subsequent editions to the "Code of Record" and addenda that are incorporated by reference in 10 CFR 50.55a(b), subject to limitations and modifications listed in 10 CFR 50.55a(b) and subject to Commission approval.

ASME Code Class 1 components include all snubbers within the reactor coolant pressure boundary. Regulatory Guide (RG) 1.26, "Quality Group Classifications and Standards for Water-, Steam-, and Radioactive-Waste-Containing Components of Nuclear Power Plants," Revision 4, dated March 2007, provides guidelines for establishing quality standards for Quality Group B, C, and D (and ASME Code classification) for water-, steam-, and radioactive-waste-containing components important to safety of water-cooled nuclear power plants, other than those in the reactor coolant pressure boundary (i.e., ASME Code Class 2 and 3 components). There are also systems of light-water-cooled reactors important to safety that are not identified in RG 1.26 for which there are established staff positions regarding quality group classification. These systems, and reference establishing their acceptable classification, are identified in Appendix A of Section 3.2.2, "System Quality Group Classification," of NUREG-800, "Standard Review Plan."

The ASME B&PV Code, Section XI and ASME OM Code are incorporated by reference in 10 CFR 50.55a(b)(3). The ASME B&PV Code, Section XI as well as the optional ASME OM Code defines the scope by stating that ISI and testing programs shall include components (including snubbers) in systems that are required to perform a specific function in (1) shutting down the reactor to a safe shutdown condition, (2) maintaining the safe shutdown condition, or (3) mitigating the consequences of an accident.

Subsection ISTA-1100 of the OM Code refers to components that are "needed to mitigate the consequences of an accident." This statement is intended to provide confidence that the health and safety of the public will be protected in the event of certain accidents and anticipated transients at a nuclear power plant. The term "accident" is also used throughout the Commission's regulations. For example, Appendix B to 10 CFR Part 50 establishes quality assurance requirements for the design, construction, and operation of "structures, systems, and components that prevent or mitigate the consequences of postulated accidents that could cause undue risk to the health and safety of the public." Similarly, 10 CFR Part 100, "Reactor Site Criteria," describes structures, systems, and components (SSCs) that must be designed to remain functional during and following a "safe shutdown earthquake" as those necessary to ensure (1) the integrity of the reactor coolant pressure boundary, (2) the capability to shut down the reactor and maintain it in a safe shutdown condition, or (3) the capability to prevent or mitigate the consequences of accidents that could result in potential offsite exposures comparable to the guideline exposures.

In establishing such requirements, the NRC uses the term "accident" to describe a broad range of possible adverse events at a nuclear power plant. Therefore, although most of the accidents of concern to inservice testing are addressed in the accident analyses chapter, licensees should be aware that the plant's final safety analysis report (FSAR) may address other accident analyses that need to be considered within the context of inservice testing.

Thus, an introductory section of the inservice inspection and testing program document submitted to the NRC for each plant must state the plant's safe-shutdown condition (i.e., hot standby, hot shutdown, cold shutdown, etc.). If the scope in Section ISTA appears to be broader than that specified in 10 CFR 50.55a, the more narrow scope applies.

Components within the scope of 10 CFR 50.55a are included in the scope of 10 CFR 50.65, "Requirements for Monitoring the Effectiveness of Maintenance at Nuclear Power Plants" (the Maintenance Rule). Licensees may elect to consolidate inspection and testing for snubbers, designating any non-Code components as such in the ISI and testing program.

The plant's FSAR (or equivalent) defines the equipment that is necessary to meet specific functions. If the FSAR indicates that a system or component is Code Class 1, 2, or 3, that system or component is within the scope of 10 CFR 50.55a. By contrast, if the FSAR states that a system or component is designed, fabricated, and maintained as Code class at the option of the Owner as permitted by Subsection ISTA-1320, the application of the related OM Code requirements is also optional.

Tables A2.1 and A2.2 (which appear at the end of this appendix) provide examples of systems with pumps and valves that licensees typically include in their inservice testing (IST) program. These tables may be used may be used for developing licensee's snubber examination and testing program. These tables are not intended to be all-inclusive, but they may form the basis for the initial review of a licensee's snubber program scope.

Figure A2.1, "Flow Chart – Development of Preservice and Inservice Examination and Testing Program for Snubbers," (which appears at the end of this chapter) provides a quick reference to regulatory requirements for development of the inservice examination and testing program for snubbers. For complete details, see 10 CFR 50.55a.

2.2.2 Snubber Attached to Steam Generator and Reactor Coolant Pumps

There are special requirements for PWR plants with regard to the testing of snubbers attached to the Steam Generators (SG) and Reactor Coolant pumps (RCP). These are generally large bore hydraulic snubbers (LBHS). Large bore hydraulic snubbers are defined as those units with rated capacities of 50 kips or greater. Unlike smaller hydraulic snubbers, LBHSs were exempt from inservice functional testing prior to 1980, primarily due to a lack of available test equipment of sufficient size. In 1980 and 1984, the NRC issued Generic Letters to all licensees requesting modification of plant TS to include LBHSs testing provisions. The results of initial tests revealed numerous cases where LBHSs were either out of specified tolerance or completely inoperable. Subsequently, the NRC developed Generic Issue (GI-113), "Dynamic Qualification and Testing of Large Bore Hydraulic Snubbers (LBHSs)," with the objective of evaluating the reliability of LBHSs in operating commercial nuclear power plants.

NUREG/CR-5416, "Technical Evaluation of Generic Issue 113: Dynamic Qualification and Testing of Large Bore Hydraulic Snubbers," dated August 1992 provided major recommendations for LBHSs. This effort was in coordination with the industry, vendors, and snubber manufacturers. As a result, the NRC established specific inservice testing recommendations for LBHSs installed on PWR steam generators or reactor coolant pumps. The NRC recommendation was that these snubbers be tested as a separate test population, and it was eventually incorporated into the ASME OM Code. Although the generic Issue focused on hydraulic snubbers, the resultant requirement does not specify a snubber size or type to which it applies. Therefore, all licensees are reminded that 10-year inservice examination and test snubber programs shall include their steam generator snubbers and reactor coolant pumps snubbers, regardless of size or type. The Code requirement is that snubbers attached to steam generators and those attached to reactor coolant pumps be designated as at least one, separate Defined Test Plan Group (DTPG) for testing purposes as specified in ISTD-5353. Large bore snubbers (greater than 50 kips) located on piping or other components may be included in the general snubber population for testing and examination purposes.

2.2.3 Testing of non-Code Snubbers

As discussed above, licensees are required to test safety-related components to demonstrate that they will perform satisfactorily in service in accordance with 10 CFR Part 50, Appendices A and B. Regulations in 10 CFR 50.55a address the inservice inspection and inservice testing program for components within the scope of the ASME Code.

An inservice examination and testing program is also a reasonable vehicle to periodically demonstrate the operational readiness of snubbers that are not covered by the Code, but are within the scope of 10 CFR Part 50, Appendices A and B. Thus, if a licensee chooses to include non-Code snubbers in its ASME Code inservice examination and testing program (or some other licensee-developed inspection and testing program) and, as a result, is unable to meet certain Code provisions for the non-Code components, the regulations (10 CFR 50.55a) do not require the licensee to submit a relief request to the NRC. Nonetheless, the licensee should maintain documentation that provides assurance of the continued operability of the non-safety components through the performed tests, and such documentation should be available for staff inspection at the plant site.

Therefore, while 10 CFR 50.55a delineates the examination and testing requirements for ASME Code Class 1, 2, and 3 snubbers, licensees should not limit their inservice inspection and testing to only those snubbers that are covered by 10 CFR 50.55a. However, care should be taken so that the inclusion of non-safety components does not adversely affect the integrity of the program by reducing the required homogenous nature of a population scope. Licensees may implement deviations from the Code for non-Code snubbers without NRC review and approval, and need not document such deviations as "relief requests." Nonetheless, a notation in the licensee's inservice examination and testing program document would help to identify the deviations and clarify that they relate to non-Code snubbers. If it is not clear that the deviations relate to non-Code snubbers, the staff might assume that the licensee is not meeting the requirements of 10 CFR 50.55a. Some licensees use the relief request format to document such deviations, while other licensees place notes, footnotes, or brief descriptions in their program documents.

2.3   Code Class Systems Containing Safety-Related Snubbers

The plant safety analysis report (SAR), technical specification (TS), and other documents list the systems and components (i.e. snubbers) that must function to support the safe operation and shutdown of the plant. Tables 2.1 and 2.2 (which appear at the end of this chapter) provide examples of systems with pumps and valves that licensees typically include in their inservice testing (IST) program for pressurized-water reactors (PWRs) and boiling-water reactors (BWRs). These tables may be used for developing licensee's snubber examination and testing program. These tables are not intended to apply to all plants; the listed systems and components are not considered safety-related at every plant, and are not necessarily classified as Code Class 1, 2, or 3. (For information on quality group and Code classifications, see RG 1.26 and Section 3.9.6 of.NUREG-0800.) The licensee's safety analysis generally contains a section describing the Code classification of components. The snubber inservice examination and testing program scope must be consistent with the SAR.

2.4   Snubber Inservice Examination and Testing Programs and their Documentation

10 CFR 50.55a(g)(4) states, in part, that throughout the service life of a boiling or pressurized water-cooled nuclear power facility, ASME Code Class 1, 2, and 3 components (including supports) meet the ISI and testing requirements of the ASME B&PV Code, Section XI or ASME OM Code as incorporated by reference in 10 CFR 50.55a(b).

2.4.1   Snubber Program while using ASME B&PV Code, Section XI

Licensees using ASME B&PV Code, Section XI for its snubber inservice examination and testing program shall consider the following rules for inservice inspection and testing of snubbers:

- Subsection IWA addresses the general requirements for inservice examination and testing of snubbers.

- Article IWF-5000 addresses inservice inspection and testing requirements for snubbers in addition to those required per Article IWF-1000.

- Subarticle IWA-1400(c) requires that owner shall submit certain plans and reports to the enforcement and regulatory authorities. Table IWA-1600-1 specifies the Edition 1987 with OMa-1988 for ASME/ANSI OM, Part 4 (OM-4) to be use while using Article IWF-5000 for snubber inservice inspection and testing.

- Article IWA-4000 provides the requirements of Repair/Replacement activities including snubbers.

- Article IWA-6000 addresses the records and reports that are required for these inspection and testing programs of snubbers.

- Subarticle IWA-6210 states that the owner shall prepare plans for preservice and inservice examinations and tests to meet the requirements of the ASME B&PV Code, Section XI requirements. Article IWF-1000 addresses the scope and responsibility for inservice examination and testing of snubbers as a subset of components support examination.

- ASME OM Code Subsections ISTA and ISTD may be used in lieu of the ASME B&PV Code, Section XI, Article IWF-5000 as allowed by the regulation 10 CFR 50.55a(b)(v)(A).

2.4.2 Snubber Program while using ASME OM Code

Licensees using ASME OM Code for its snubber inservice examination and testing program shall consider the following rules for inservice inspection and testing of snubbers. Subsection ISTA includes general requirements (including scope) for inservice examination and testing of snubbers.

- Subsection ISTD addresses the "Preservice and Inservice Examination and Testing of Dynamic Restraints (Snubbers) in Light-Water Reactor Nuclear Power Plants."

- Subsection ISTA-3200 states that IST plans shall be filed with the regulatory authorities. This includes snubber examination and testing plans.

- Subsection ISTA-9000 addresses the records and reports that are required for these inspection and testing programs.

- Subsection ISTA-9210 states that the owner shall prepare plans for preservice and inservice examinations and tests to meet the requirements of the OM Code.

- Subsection ISTA-9220 states that licensees shall prepare examination, and test records in accordance with the requirements of this article in conjunction with the applicable Subsection of the OM Code (i.e. ISTD).

- Nonmandatory Appendix A, and the Supplements to Nonmandatory Appendix A to the OM Code describe voluntary guidance for licensees to use in preparing their inspection and test plans.

2.4.3 Snubber Program while using NRC Authorized Alternative or Relief

Licensees not using the ASME Section XI or ASME OM requirements for their snubber program, must submit a request for relief from or an alternative to the ASME Section XI and ASME OM requirements. NRC-authorized alternatives to use TRMs or other-licensee-controlled documents, in lieu of the ASME B&PV Code, Section XI or ASME OM Code requirements for inservice examination and testing of snubbers, do not preclude the need for licensees to submit snubber examination and test plans to the regulatory authorities as defined in IWA-1400(c) and ISTA-3200. An NRC-authorized alternative is only applicable to the requested Sections or paragraphs of the ASME Code, not from the entire ASME Code.

2.4.4 Snubber Programs and Their Bases

Some licensees are using TRMs or other licensee-controlled documents for snubber inservice examination and testing in lieu of the ASME B&PV Code, Section XI requirements. TRMs or other licensee-controlled documents serve as bases for most snubber programs and most of the snubber programs have similarities across the industry. Many licensees are in the process of updating their snubber programs to comply with and incorporate the ASME OM Code. Some licensees have already updated their programs to use ASME OM Code. As a minimum, the updated snubber program documentation shall contain sufficient information to verify alignment with the ASME OM Code requirements. Bases documents have typically included a description of the methodology used in preparing the snubber program plans. The bases document shall clearly state where and how a list of program snubbers is kept and maintained. Although not required by the regulation, the bases documents will help licensees ensure the consistent implementation of their snubber programs throughout the course of typical organizational (including personal) changes. A good bases document will also enable the plant staff to clearly understand the snubber categorization process, as well as the basis for examination and testing requirements. The bases document can also serve as a useful reference for reviews performed under 10 CFR 50.59 when changes are made to a facility.

As a minimum, the following three elements as described in the OM Code are recommended to be addressed in a typical snubber program bases document:

(1)   Visual Examination Requirements

(2)   Functional Testing Requirements

(3)   Service Life Monitoring Requirements

Individual aspects of each element shall be detailed as outlined in the following sections in order to provide clear definitions and descriptions of the program elements. Information clarifying the basis for inclusion in the program shall be provided, including references to applicable Code sections or licensing commitments where appropriate.

To help ensure consistency throughout the industry, licensees are encouraged to use these guidelines and to consult with the Snubber User Group (SNUG) for guidance when developing snubber programs and their bases.

2.4.4.1 Inservice Visual Examination (description, definitions, and basis of each item)

(1) Addressing integral and nonintegral attachments to snubbers (see Section 2.8)

(2) Snubber population groupings

(3) Initial Examination Intervals

(4) Subsequent Examination Intervals

(5) Method of Visual Examination

(6) Inservice Examination Failure Evaluation

(7) NRC authorized alternative (or relief request), if applicable

(8) Code Case used for visual examination, if applicable

(9) Corrective action plans

2.4.4.2 Inservice Operational Readiness or Functional Test (description, definitions, and basis of each item)

(1) Functional Test Frequency (every Refueling Outage)

(2) Test Plan Groups [Defined Test Plan Group (DTPG)]

(3) Sample Plans (10% testing sample, 37 testing sample, or 55 testing sample) used and Initial Snubber Sample size(s) anticipated for each DTPG (It is recognized that 10% Plan samples may vary slightly over the course of an interval due to ongoing station modifications or replacement. The bases document may be periodically updated for significant changes but is not expected to be a day-to-day "living" document.)

(4) Additional Snubber Sampling Method for each plan (based on selected sample plan on 10% testing plan, 37 testing sample or 55 testing sample)

(5) Failure Evaluation requirements and methods (Reference applicable document)

(6) Test Failure Mode Groupings methodology (Reference applicable document)

(7) Corrective Actions for each sample plan and FMG identified (Reference applicable document)

(8)   NRC authorized alternative (or relief request), if applicable

(9)   Code Case used for functional testing of snubbers, if applicable

2.4.4.3 Service Life Monitoring Program

The licensees must develop the service-life monitoring (SLM) program as defined in the ASME OM Code, Subsection ISTD-6000 or TS or TRM or in accordance with an approved Relief Request. SLM must include all the maintenance record data available for snubbers while evaluating or reevaluating the service life. Nonmandatory Appendix F of ISTD provides additional guidance in developing a SLM Program. The SLM program is the primary instrument for assuring continued reliability of a snubber population at a plant. The statistical method sample testing is point-in-time assessment of population functionality but in general does not serve as an effective tool to either maintain or improve reliability. This is due to the fact that such functional testing is based on small samples (10% or 37 snubbers) every refueling outage, and is not predictive in nature. Based on snubber aging study information, in the NUREG/CR-5870, "Results of LWR Snubber Aging Research," dated May 1992, the NRC recommended the inclusion of SLM of snubbers in addition to the statistical testing process. Most licensees have included some reference to SLM in their existing programs. The ASME OM Code, Subsection ISTD also included SLM along with snubber examination and inservice testing requirements. Some of the licensees have updated their snubber programs to incorporate ASME OM Code requirements. The updated snubber programs often simply reference plant procedures for snubber examinations and testing without providing any references to applicable sections of the ASME OM Code. Program documentation shall provide information regarding specific SLM requirements and how the requirements are satisfied.

The records of all activities (i.e repair, replacements, maintenance, corrective action work, failures, etc.) related to all snubbers must be available and considered for the SLM program.

2.4.5   Snubber List or Snubber Controlled Data Bases

In preparing and maintaining a snubber list or data base, licensees shall consider the ability to produce reports providing adequate information to both implement and assess the program. Reports generated to provide a snubber listing should include the following suggested headings, which are shown along with a description of the information that licensees might produce under each heading.

Title: Report name, including the applicable plant and unit.

Reference information: Include references to maintaining and location of controlled snubber data. This may be a controlled station data base from which reports and lists are generated as needed.

Program/Report revision or revision date: List the revision number and date/or date (on each page).

System, Code class, and group: List applicable information such as plant system, ASME Code Class, and type of snubber (hydraulic or mechanical).

Snubber identification: List a unique identifier for each snubber; this identifier shall be used consistently in all snubbers' inservice examination and testing program documentation and design information such as system piping and instrument diagrams (P&IDs), isometric, test procedures, and relief requests.

Drawings number: List the applicable isometrics, support drawings or figures that depict the snubber.

The following items may not normally be included in a list of individual snubbers, as they typically apply to entire populations. This information shall be included as annotations for any individual snubbers for which a specific frequency or relief request applies on a unique basis.

Test frequency: List the actual frequency for each inspection and test to be performed.

Relief request(s): List any applicable relief requests in the snubber list.

2.4.6   Snubber Program Plan Documentation and Their Submittal to NRC

10 CFR 50.55a(g)(4) requires that, throughout the service life of a boiling or pressurized water-cooled nuclear power facility, ASME Code Class 1, 2, and 3 components (including supports) meet the requirements of the ASME B&PV Code, Section XI or ASME OM Code as incorporated by reference in 10 CFR 50.55a(b). The applicable ASME B&PV Code, Section XI, Article IWA-1000, "General Requirements," and ASME OM Code, Subsection ISTA, "General Requirements," provide the documentation and submittal requirements for inservice testing and examination of certain components in light-water nuclear power plants. Therefore, based on these requirements, licensees are required to submit their snubber examination and testing plans and their updates every 120 months.

(a)   Documentation requirements for snubber program plan when using the ASME B&PV Code, Section XI

IWA-1400(c) notes that owners have the responsibility to prepare plans, schedules, and inservice inspection summary reports, and submit of these plans and reports to the enforcement and regulatory authorizes having jurisdiction at the plant site.

Article IWA-6000, "Record and Reports," provides the requirements for preparation, submittal, and retention of records and reports.

(b)   Documentation requirements for snubber program plan when using the ASME OM Code

ISTA-3200(a) requires that plans for inservice examination and testing of snubbers shall be filed with the regulatory authorities having jurisdiction at the plant site.

ISTA-9000, "Records and Reports," provides the requirements for preparation, submittal, and retention of records and reports.

Nonmandatory Appendix-A and the Supplement to Nonmandatory Appendix-A describe voluntary guidance for licensees to develop snubber inservice examination and testing plans.

(c)   Documentation requirements for snubber programs when using NRC authorized alternative TS, TRM or other licensee controlled documents in lieu of the ASME B&PV Code, Section XI, or ASME OM Code

NRC-authorized relief to use TRMs or other-licensee-controlled documents, in lieu of the ASME B&PV Code, Section XI or ASME OM Code requirements for inservice examination and testing of snubbers, do not provide relief from submitting snubber examination and test plans and reports to the regulatory authorities. Submittal is required by the applicable ASME B&PV Code, Section XI or ASME OM Code as noted in (a) and (b) above. Licensees not meeting the requirements of IWA-1400(c) or ISTA-3200(a) must submit appropriate documents containing snubber inservice examination and testing plans and submit a request for relief to the NRC pursuant to 10 CFR 50.55a(a)(3). NRC staff generally not perform a review of submitted snubber inservice examination and testing plans and reports unless the applicant or licensee requests alternatives or reliefs to Code requirements.

2.5   Relief Requests and Proposed Alternatives

Licensees are required to perform the inservice examination and testing of snubbers in accordance with ASME B&PV Code, Section XI or the OM Code and the applicable addenda as required by 10 CFR 50.55a(g) or 10 CFR 50.55a(b)(3)(v), except where the NRC has granted specific written relief, pursuant to 10 CFR 50.55a(g)(6)(i), or authorized alternatives pursuant to 10 CFR 50.55a(3). 10 CFR 50.55a(a)(3) states that licensees may use alternatives to the requirements of 10 CFR 50.55a(g) when authorized by the NRC if (1) the proposed alternatives would provide an acceptable level of quality and safety, or (2) compliance with the specified requirements would result in hardship or unusual difficulty without a compensating increase in the level of quality and safety. A licensee may submit a request for the NRC to review and approve relief from requirements of the Code, or to authorize the use of proposed alternatives.

The justification must include adequate information for the staff to determine if the relief can be granted or the alternative can be authorized. NRC approval is required before a licensee may implement proposed alternatives that must be authorized pursuant to 10 CFR 50.55a(3). By contrast, a licensee may implement proposed alternative testing while the NRC is reviewing requests for relief from Code requirements made pursuant to 10 CFR 50.55a(f)(6)(i), if the licensee has determined that the requirements are impractical.

The staff performs a detailed review of each relief request, grants relief from the requirements or authorizes an alternative to those requirements, and may impose alternative requirements. When granting relief, the NRC considers the burden that would be imposed upon the licensee if the agency enforced the specified requirements.

For more details contents and format, etc. see Nuclear Energy Institute (NEI) issued document, "Standard Format for Requests from Commercial Reactor Licensees Pursuant to 10 CFR

50.55a," Revision 1, dated 2004. Currently, few licensees are using the ASME OM Code to meet the requirements of 10 CFR 50.55a for snubber inservice examination and testing, whereas most of the licensees are using a variety of licensee-controlled documents or procedures in lieu of the applicable ASME Code requirements. These licensee-controlled documents or procedures include the following:

(1) Technical Specification (TS)
(2) Technical Requirement Manual (TRM)
(3) Final Safety Analysis Report (FSAR)
(4) Updated Final Analysis Report (UFSAR)
(5) Selected Licensee Commitment (SLC)
(6) Licensee-Controlled Specification (LCS)
(7) Equipment Control Guidelines (ECG)
(8) Other Licensee-Controlled Procedures.

Recently, the NRC staff has identified several instances in which nuclear power plants licensees have used a TRM, or other licensee-controlled documents and procedures, which do not meet requirements of their "Code of Record" for the ISI and testing of snubbers. These licensees have not requested approval to use these alternatives from the NRC. The NRC issued Regulatory Issue Summary (RIS) 2010-06, "Inservice Inspection and Testing Requirements of Dynamic Restraints (Snubbers)" on June 1, 2010, to inform licensees of the NRC's rules and regulations regarding snubber ISI and testing, in accordance with 10 CFR 50.55a(g), at nuclear power plants.

In addition, it is noted that once relief has been granted all documents included in or referenced by the relief request become equivalent to licensing commitments and cannot be changed in a substantive manner without resubmitting the request for approval. It may be inappropriate to make substantive changes to licensing documents under the 10 CFR 50.59 process. Any changes to TRMs or other referenced documents must be evaluated in view of the original relief request and assessed as to the appropriate change process.

The NRC expects licensees to ensure that their snubber ISI and testing programs are in compliance with 10 CFR 55.55a(g) or authorized alternatives. If licensees discover that their programs are not meeting 10 CFR 50.55a(g) requirements or authorized alternatives, they shall take appropriate actions to bring their programs back into compliance and ensure that non-compliant systems, structures and components are operable. In certain circumstances involving snubber programs at nuclear power plants that are not in compliance with NRC requirements, enforcement discretion has been provided by the NRC. The NRC's Office of Enforcement issued Enforcement Guidance Memorandum (EGM)-10-001, "Dispositioning Violation of Inservice Examination and Testing Requirements for Dynamic Restraints (Snubbers)," on June 1, 2010, to provide NRC staff guidance for the disposition of certain 10 CFR 50.55a violations and the potential of granting enforcement discretion for the affected requirements. The NRC expects that licensees of nuclear power plants, who were not meeting the 10 CFR 50.55a requirements for snubber inservice examination and testing as described in RIS 2010-06, shall have entered any noncompliance into their corrective action program and corrected the noncompliance by meeting the applicable ASME Code requirements or by submitting for relief to the NRC.

2.6   Snubber Program Plan and its Update Documents

Snubber inservice examination and testing program plans submitted to the NRC are used to prepare for NRC inspections and to address other licensing actions that may arise. To facilitate these regulatory activities, the NRC would like to receive up-to-date program plan documents when the licensee makes significant changes to the snubber inservice examination and testing program plan in the interim period between the required 10-year interval plan submittals. These interim informational submittals are generally considered "good faith" and not a regulatory requirement. As long as the snubber inservice examination testing program plan remains consistent with the regulations, ASME Code relief is not required for these interim updates. That is, deletions from or additions to the snubber program do not necessarily require NRC approval, unless commitments to obtain such approval exist as a result of prior approved relief requests or similar commitments. The burden is on each licensee to verify that its snubber program is complete and includes all snubbers that require inservice examination and testing. If a licensee deletes a particular snubber from its snubber program plan, the staff recommends that the licensee should document the basis in an appropriate manner. Such changes do not require approval, unless cumulative changes affect the examination or testing plans in a significant manner. Changes in sample plans and DTPG groupings require notification of the regulator, but do not require approval.

The staff expects each licensee to maintain its snubber examination and testing program plan up-to-date and ensure that it remains consistent with changes in plant configuration. Conversely, if a system modification results in the addition of a snubber to the snubber program plan, the licensee shall ensure that it is incorporated into the program to satisfy the Code and licensing requirements or that a relief request is submitted for NRC review and approval, as appropriate.

2.7   Repair and Replacement of Snubbers

The repair and replacement of snubbers is to be performed using applicable edition and addenda of the ASME B&PV Code, Section XI, Article IWA-4000. The NRC-approved Inservice Inspection Code Cases in RG 1.147, such as N-508-3, "Rotation of Serviced Snubbers and Pressure Relief Valves for the purpose of Testing, Section XI, Division 1," may be used.

2.8   ISI of the Integral and Non-Integral Attachments of Supports Containing Snubbers

Inservice inspection of integral and non-integral attachments, such as lugs, bolts, pins, and clamps must be performed by use of ASME B&PV Code, Section XI, Subsection IWF or other applicable Subsections. ASME OM Code, Subsection ISTD covers inservice examination and testing of snubbers (pin to-pin) inclusive and does not address the inspection of integral and non-integral attachments, such as lugs, bolting, pins, and clamps.

2.9   Developing Snubber Program for New Nuclear Power Plants

Under 10 CFR Part 52, "Licenses, Certifications, and Approvals for Nuclear Power Plants," the development of a plant-specific IST program (or snubber examination and testing program) is the responsibility of the applicant for a combined license (COL) to construct and operate a nuclear power plant. The Commission's Staff Requirements Memorandum (SRM), dated September 11, 2002, for Commission Paper SECY-02-0067, "Inspections, Tests, Analyses, and Acceptance Criteria (ITAAC) for Operational Programs (Programmatic ITAAC)," stated that ITAAC for an operational program are unnecessary if the program and its implementation are fully described in the COL application and found to be acceptable by the NRC. The Commission also stated that the burden is on the COL applicant to provide the necessary and sufficient programmatic information for approval of the COL without ITAAC. In its May 14, 2004, SRM for SECY-04-0032, "Programmatic Information Needed for Approval of a Combined License Without Inspections, Tests, Analyses and Acceptance Criteria," the Commission defined "fully described" as meaning that the program is clearly and sufficiently described in terms of the scope and level of detail to allow a reasonable assurance finding of acceptability. The Commission also noted that required programs should always be described at a functional level and at an increasing level of detail where implementation choices could materially and negatively affect the program effectiveness and acceptability. SECY-05-0197, "Review of Operational Programs in a Combined License Application and Generic Emergency Planning Inspections, Tests, Analyses, and Acceptance Criteria," summarizes the NRC position regarding the full description of operational programs to be provided by COL applicants. RG 1.206, "Combined License Applications for Nuclear Power Plants (LWR Edition)," provides guidance for COL applicants with respect to fully describing plant operational programs.

COL applicants provide a description of the snubber examination and testing program for NRC review as part of their COL applications. In some cases, the COL applicant incorporates by reference in its Final Safety Analysis Report (FSAR) the description of the snubber examination and testing program provided in the Design Certification documentation, such as the Design Control Document (DCD) or Design Certification FSAR. RG 1.206 provides guidance regarding information to be provided in the COL application as part of the description of the inservice examination and testing program for snubbers. To date, COL applicants have described their snubber examination and testing program based on the requirements in ASME OM Code, Subsection ISTD. The guidance in this NUREG may be used in developing and implementing the IST program including snubber inservice examination and testing program for new nuclear power plants.

The NRC staff is reviewing the descriptions of IST programs, including snubber examination and testing programs, provided by COL applicants as part of their COL applications. The staff will reach a conclusion regarding the acceptability of the description of the IST program in its safety evaluation on the COL application. The NRC staff will conduct inspections of the development and implementation of IST programs during the construction of new nuclear power plants following COL issuance.

2.10   Technical Specification (TS) Improvement To Modify Requirements Regarding The Addition of Limiting Condition for Operation (LCO) 3.0.8 on the Inoperability of Snubbers

The NRC staff issued a *Federal Register* Notice (69 FR 68412) on November 24, 2004, that provided a Model Safety Evaluation and Model Application related to the addition of limiting condition for operation (LCO) 3.0.8 for inoperable snubbers on supported systems in technical specifications (TS). The purpose of this model is to permit the NRC to efficiently process amendments that propose to modify requirements by adding LCO 3.0.8 to TS to provide a delay time for entering a supported system TS when the inoperability is due solely to an inoperable snubber, provided risk is assessed and managed, as generically approved by this notice. Licensees of nuclear power reactors to which the model applies could request amendments utilizing the model application.

**Table A2.1 Typical Systems or Portions of Systems in the Scope of 10 CFR 50.55a
Where Snubbers are included in the
Inservice Examination and Testing Snubber Program
for a Pressurized-Water Reactor (Non-Inclusive)**

Typical safety-related, Code-class system in pressurized-water reactors
Reactor coolant system and flowpaths for establishing natural circulation
Main steam system
High-pressure safety injection system
Chemical and volume control or makeup system
Low-pressure safety injection system
Shutdown cooling, residual heat removal, or decay heat removal systems
Containment spray system
Main feedwater system
Auxiliary feedwater system
Primary containment system
Component cooling water system
Spent fuel pool/pit cooling system
Service water system
Emergency diesel generator system (within scope of 10 CFR 50.55a)
Ventilation systems
Instrument air system (if within the scope of 10 CFR 50.55a)

Table A2.2   Typical Systems or Portions of Systems in the Scope of 10 CFR 50.55a
Where Snubbers are included in the
Inservice Examination and Testing Snubber Program
for a Boiling-Water Reactor (Non-Inclusive)

Typical safety-related, Code-class system in boiling-water reactors
Nuclear boiler and reactor recirculation system
Main steam system
High-pressure core coolant injection (HPCI) system
High-pressure core spray system
Reactor core isolation cooling (RCIC) system (if safety-related)
Reactor water cleanup system
Residual heat removal (RHR) system
Spent fuel pool cooling system
Feedwater coolant injection and isolation condenser system (if applicable)
Standby liquid control (SBLC) system
Main feedwater system
Primary containment system
Closed cooling or component cooling water system
Service water system
Control rod drive system (portions within the scope of 10 CFR 50.55a)
Emergency diesel generator systems (if within the scope of 10 CFR 50.55a)
Ventilation systems
Instrument air system (if within the scope of 10 CFR 50.55a)
Traversing incore probe system (if within the scope of 10 CFR 50.55a)

Figure A2.1 – Appendix-A

Flowchart - Development of Preservice and Inservice Inservice Examination and Testing Program for Snubbers*

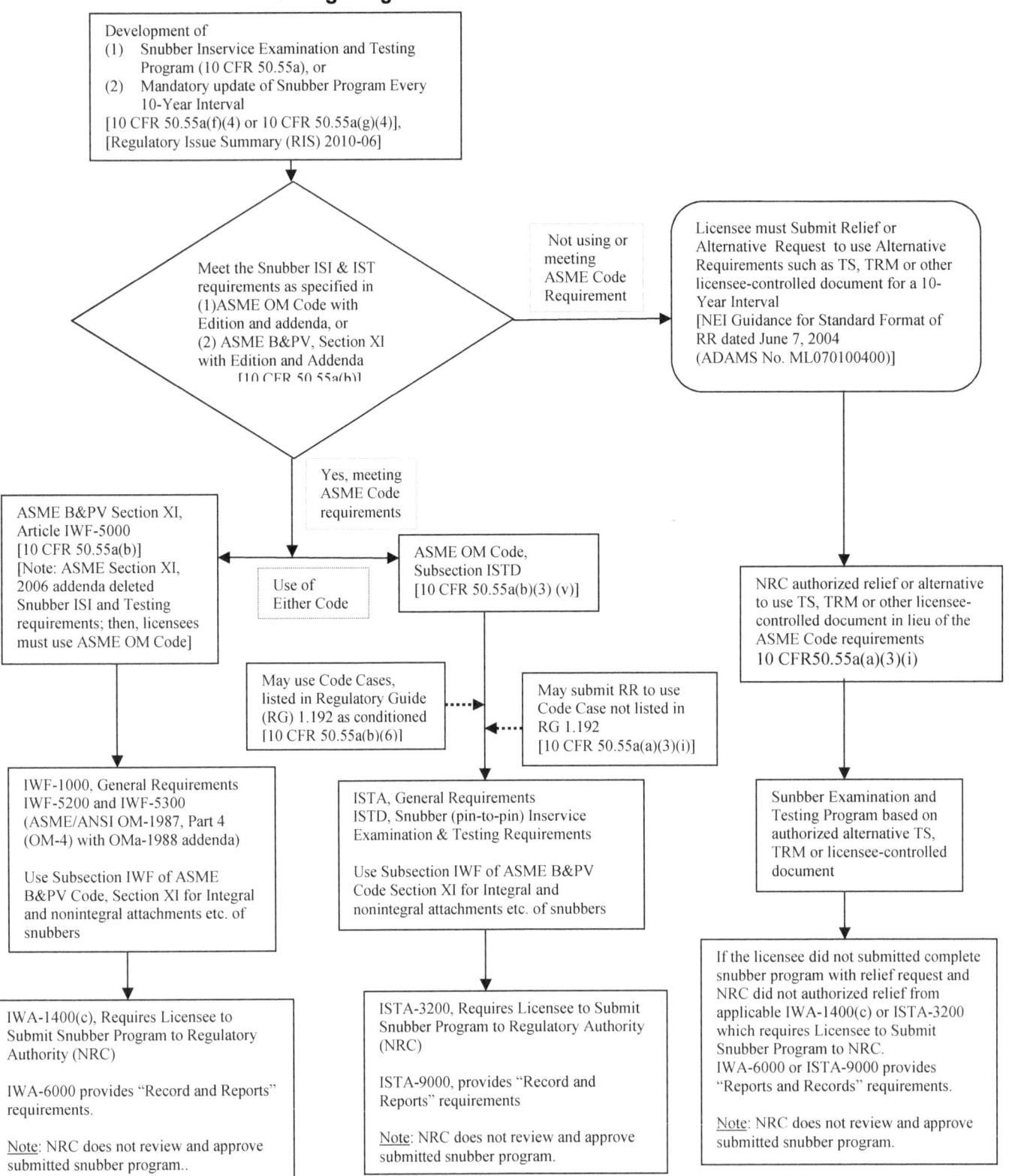

*Note: Flow chart provided for guidance only. For complete details see 10 CFR 50.55a

3. REFERENCES

3.1   *U.S. Code of Federal Regulations*, Domestic Licensing of Production and Utilization Facilities, Title 10, "Energy," Chapter I – Nuclear Regulatory Commission, Part 50, Section 50.55a, Codes and standards.

3.2   *American Society of Mechanical Engineers Boiler and Pressure Vessel Code,* Section XI, "Rules for Inservice Inspection of Nuclear Power Plant Components," 2004 Edition thru 2008 Addenda.

3.3   *American Society of Mechanical Engineers Operation and Maintenance Code,* "Rules for Inservice Testing of Light-Water Reactor Power Plants." 2004 Edition with 2005 and 2006 Addendas

3.4   *American Society of Mechanical Engineers Operation and Maintenance Code,* Operation and Maintenance of Nuclear Power Plants, Part 4 (OM-4) 1987 Edition with 1988 Addenda.

3.5   U.S. Nuclear Regulatory Commission, Regulatory Guide 1.192, "Operation and Maintenance Code Case Acceptability: ASME OM Code," Washington, DC, June 2003.

3.6   U.S. Nuclear Regulatory Commission, Regulatory Guide 1.193, "ASME Code Cases Not Approved for Use," Washington, DC, June 2003.

3.7   U.S. Nuclear Regulatory Commission, Regulatory Guide 1.147, "Inservice Inspection Code Case Acceptability, ASME Section XI, Division 1," Washington, DC, October 2010.

3.8   U. S. Nuclear Regulatory Commission, Regulatory Guide 1.187, "Guidance for Implementation of 10 CFR 50.59, Changes, Tests, and Experiments," Washington, DC, November 2000.

3.9   U. S. Nuclear Regulatory Commission, Regulatory Guide 1.206, "Combined Licensee Application for Nuclear Power Plants (LWR Edition)," June 2007.

3.10   U. S. Nuclear Regulatory Commission, Regulatory Issue Summary 2004-12, "Clarification on use of Later Editions and Addenda to the ASME OM Code and Section XI."

3.11   U. S. Nuclear Regulatory Commission, Regulatory Issue Summary 2012-08, "Developing Inservice Testing and Inservice Inspection Programs Under 10 CFR Part 52."

3.12   U.S. Nuclear Regulatory Commission, Generic Letter (GL) 90-09, "Alternative Requirements for Snubber Visual Inspection Intervals and Corrective Actions," December 11, 1990.

3.13   Nuclear Energy Institute (NEI), White Paper, "Standard Format for Requests from Commercial Reactor Licensees Pursuant to 10 CFR 50.55a, Revision 1," June 2004, ADAMS Accession No. ML070100400.

3.14 U.S. Nuclear Regulatory Commission, Regulatory Issue Summary (RIS) 2010-06, "Inservice Inspection and Testing Requirements of Dynamic Restraints (Snubbers)," June 1, 2010.

3.15 U.S. Nuclear Regulatory Commission, Enforcement Guidance Memorandum (EGM)-10-001, "Dispositioning Violation of Inservice Examination and Testing Requirements for Dynamic Restraints (Snubbers)," June 1, 2010.

3.16 U.S. Nuclear Regulatory Commission, Generic Issue (GI-113), "Dynamic Qualification and Testing of Large Bore Hydraulic Snubbers (LBHSs)," NUREG-0933, Main Report with Supplement 1-34.

3.17 U.S. Nuclear Regulatory Commission, "Results of LWR Snubber Aging Research," NUREG/CR-5870, May 1992, ADAMS Accession No. 9206120267.

4. HISTORICAL REFERENCES

4.1 U.S. Nuclear Regulatory Commission, (NRC) Generic Letters:

    4.1.1   GL 84-13, "Technical Specification For Snubbers"

    4.1.2   GL 89-09, "ASME III Component Replacement"

    4.1.3   GL 90-09, "Alternative Requirements for Snubber Visual Inspection Intervals and Corrective Actions"

4.2 NRC Inspection Enforcement (IE) Bulletins

    4.2.1   BL-73-03, "Defective Hydraulic Shock Suppressors and Restraints"

    4.2.2   BL-73-04, "Defective Bergen-Patterson Hydraulic Shock Suppressors Absorbers"

    4.2.3   BL-75-05, "Operability of Category I Hydraulic Shock and Sway Suppressors"

    4.2.4   BL-78-10, "Bergen-Patterson Hydraulic Shock Suppressor Accumulator Spring Coil"

    4.2.5   BL-79-02, "Pipe Support Base Plate Designs Using Concrete Expansion Anchors Bolts"

    4.2.6   BL-79-14, Revision 1, Supplement 2, "Seismic Analysis for As-Build Safety-Related Piping"

    4.2.7   BL-81-01, Revision 1, "Surveillance of Mechanical Snubbers"

4.3 NRC Information Notices (IN):

    4.3.1   IN 79-01, "Bergen-Patterson Hydraulic Shock Arrestors"

    4.3.2   IN 79-05, "Improper Material in Safety-Related Components"

    4.3.3   IN 83-13, "Design Misapplication of Bergen-Patterson Resistance Clamp"

    4.3.4   IN 83-20, "ITT Grinnell Figure 306/307 Mechanical Snubber Attachment Interference"

    4.3.5   IN 83-47, "Failure of Hydraulic Snubbers as a Result on Contaminated Hydraulic Fluid"

    4.3.6   IN 84-67, "Recent Snubber Inservice Testing with High failure Rates"

    4.3.7   IN 84-73, "Down Rating of Self-Aligning Ball Bushing in Snubbers"

4.3.8 IN 86-102, "Repeated Multiple Failure of Steam Generator Hydraulic Snubbers due to Control Valve Sensitivity"

4.3.9 IN 89-30, "High Temperature Environment at Nuclear Power Plants"

4.3.10 IN 94-48, " Snubber Lubrication Degradation in High Temperature Environments"

4.3.11 IN 95-09, "Use of Inappropriate Guidelines and Criteria for Nuclear Piping and Pipe Support Evaluation and Design"

4.3.12 IN 97-16, "Preconditioning of Plant Structures, Systems, and Components before ASME Code Inservice Testing or Technical Specification Surveillance Testing"

4.4 NRC IE Circulars (Cr):

4.4.1 Cr 76-05, "Hydraulic Shock and Sway Suppressors – Maintenance of Bleed and Lock-Up Velocities on ITT Grinnell's Model Nos. Figure 200 and Figure 201, Catalog PH-74-R"C 78-07, " Damaged Components of Bergen Paterson Series 25000 Test Stand"

4.4.2 Cr 79-25, "Shock Arrestor Strut Assembly Interference"

4.4.3 Cr 81-05, "Self-Aligning Rod End Bushing for Pipe Supports"

4.5 NUREGs:

4.5.1 NUREG/CR-5386, "Basis for Snubber Aging Research: Nuclear Plant Aging Research Program, Vol. 1," dated January 1990," ADAMS Accession No. ML040360184.

4.5.2 NUREG/CR-4279, "Aging and Service Wear of Hydraulic and Mechanical Snubbers Used Safety-Related Piping and Components of Nuclear Power Plant - Phase I Study," dated February 1986, ADAMS Accession No. ML040230419.

4.5.3 NUREG/CR-5870, "Results of LWR Snubbers Aging Research," dated May 1992, ADAMS Accession No. ML040340438.

4.6 Other Snubbers Related Documents:

4.6.1 PNL-SA-20219, "ASME Subsection ISTD Recommendation Based upon NAPR Snubber aging Research Results," Dated December 1991, Pacific Northwest Laboratory, Richland, WA-99352, (Work Supported by U. S. Department of Energy), ADAMS Accession No. ML04290285.

4.6.2 Resolution of Generic Safety Issues: Issue 113: Dynamic Qualification Testing of Large Bore Hydraulic Snubbers (Rev-2) (NUREG-0933, Main Report with Supplements 1-34.

NRC FORM 335 (12-2010) NRCMD 3.7	U.S. NUCLEAR REGULATORY COMMISSION BIBLIOGRAPHIC DATA SHEET *(See instructions on the reverse)*	1. REPORT NUMBER (Assigned by NRC, Add Vol., Supp., Rev., and Addendum Numbers, if any.) NUREG-1482, Revision 2 Final

2. TITLE AND SUBTITLE Guidelines for Inservice Testing at Nuclear Power Plants Inservice Testing of Pumps and Valves and Inservice Examination and Testing of Dynamic Restraints (Snubbers) at Nuclear Power Plants	3. DATE REPORT PUBLISHED	
	MONTH	YEAR
	October	2013
	4. FIN OR GRANT NUMBER	

5. AUTHOR(S) Gurjendra S. Bedi	6. TYPE OF REPORT Technical
	7. PERIOD COVERED *(Inclusive Dates)*

8. PERFORMING ORGANIZATION - NAME AND ADDRESS (If NRC, provide Division, Office or Region, U. S. Nuclear Regulatory Commission, and mailing address; if contractor, provide name and mailing address.)
Division of Engineering
Office of Nuclear Reactor Regulation
U. S. Nuclear Regulatory Commission
Washington, DC 20555-0001

9. SPONSORING ORGANIZATION - NAME AND ADDRESS (If NRC, type "Same as above", if contractor, provide NRC Division, Office or Region, U. S. Nuclear Regulatory Commission, and mailing address.)

Same as above

10. SUPPLEMENTARY NOTES
Gurjendra S. Bedi, Project Manager

11. ABSTRACT (200 words or less)

In Revision 2 of NUREG-1482, the staff of the U. S. Nuclear Regulatory Commission (NRC) discusses the applicable regulations for the inservice testing of pumps and valves, and the examination and testing of dynamic restraints (snubbers) at commercial nuclear power plants. The information in NUREG-1482, "Guidelines for Inservice Testing at Nuclear Power Plants," Revision 0, issued April 1995, and Revision 1, issued January 2005, has described this topic in the past. This NUREG report replaces Revision 0 and Revision 1 to NUREG-1482, and is applicable, unless stated otherwise, all editions and addenda of the American Society of Mechanical Engineers Code of Operation and Maintenance of Nuclear Power Plants (OM Code), and ASME Boiler & Pressure Vessel Code (B&PV), which Titles 10 of the Code of Federal Regulation (10 CFR) 50.55a(b) incorporated by reference (Federal Register, Vol. 76, No. 119, page 36232-36279, dated June 21, 2011). This NUREG-1482, Revision 2 incorporates all the public comments received against draft NUREG-1482, Revision 2 (ADAMS Accession Number: ML112231412). Based on public comments, all the structure and sections of NUREG-1482, Revision 1 are maintained or revised, and added new Appendix-A for guidance for developing and implementing an inservice examination and testing program for dynamic restraints (snubbers). In addition, the NUREG discusses other inservice test and examination program topics such as the NRC process for review of the OM Code conditions on the use of the OM Code, and interpretation of the OM Code.

12. KEY WORDS/DESCRIPTORS (List words or phrases that will assist researchers in locating the report.) NUREG-1482 — Valves Inservice Testing — Pumps IST — Valve Testing 10 CFR 50.55a — Pump Testing ASME OM Code — Dynamic Restraints ASME B&PV — Snubbers ASME Section XI — Examination Relief Requests — Snubber Visual Examination Alternative — Snubber Testing	13. AVAILABILITY STATEMENT unlimited
	14. SECURITY CLASSIFICATION
	(This Page) unclassified
	(This Report) unclassified
	15. NUMBER OF PAGES
	16. PRICE

NRC FORM 335 (12-2010)

Printed
on recycled
paper

Federal Recycling Program

UNITED STATES
NUCLEAR REGULATORY COMMISSION
WASHINGTON, DC 20555-0001

————

OFFICIAL BUSINESS

NUREG-1482, Rev. 2
Final

Guidelines for Inservice Testing at Nuclear Power Plants

October 2013

www.ingramcontent.com/pod-product-compliance
Lightning Source LLC
Chambersburg PA
CBHW080242180526
45167CB00006B/2382